T0201980

Tesi di perfezionamento in Fisica sostenuta il 16 ottobre 1997

Sergio Conti
Max-Planck-Institut für Mathematik
in den Naturwissenschaften
Inselstraße, 22-26
D-04103 LEIPZIG
Germany
e-mail: sergio.conti@mis.mpg.de

Ground state properties and excitation spectrum of correlated electron systems

PACS (Physics and Astronomy Classification Scheme): 71.45.Gm, 73.20.Dx, 73.20.Mf

Sergio Conti

Ground state properties
and excitation spectrum
of correlated electron systems

TESI DI PERFEZIONAMENTO

SCUOLA NORMALE SUPERIORE
1999

ISBN: 978-88-7642-269-0

Contents

Introduction

Electron gas theory is one of the broadest fields in theoretical condensed matter physics, and even its most elementary application to the study of collective excitations and screening in the simple metals still poses interesting questions. Recent Electron Energy Loss and Inelastic X–Ray Scattering experiments have shown that traditional electron gas theories are unable to account for the measured plasmon dispersion relation. While it has become clear that neither correlation nor band structure alone can explain those results, the recently developed Time Dependent Density Functional Theory (TD–DFT) provides a general framework – in principle, exact – which can account for both. TD–DFT allows to compute dynamical properties of inhomogeneous interacting systems, such as optical spectra and collective excitation frequencies, and relies on the presence of practical approximation schemes for the dynamic exchange and correlation potential, which is a universal functional of the time–dependent density. The generality of the method motivates strong interest in the derivation of sensible approximations, and the relevance of dynamical phenomena prevents wide applicability of the simple adiabatic *Ansatz*. Detailed analysis of constraints deriving from basic conservation laws has shown that most of the dynamical approximations suggested in the past are inconsistent, and moreover that no consistent fully dynamical local approximation to TD–DFT can be obtained within the traditional scheme based on the time–dependent density alone.

Vignale and Kohn have shown that such problems can be avoided by resorting to Current Density Functional Theory (CDFT), and furthermore that in the linear response regime and for spatially slowly varying densities the dynamical exchange and correlation potential can be exactly expressed in terms of the exchange and correlation kernels for the homogeneous electron gas. Their result has afterwards been extended to the nonlinear regime, within a generalized hydrodynamical framework, and provides a general scheme for application of TD–CDFT in the long wavelength limit. Practical implementations have so far been lacking, partly due to the absence of reliable estimates for the frequency–dependent long–wavelength exchange–correlation kernels of the homogeneous electron gas, which coincide with the customary local field corrections of electron-gas theory. A number of approximate schemes have been proposed which study the local field correction as a function of wavevec-

tor alone, neglecting its frequency dependence, and recently exact Quantum
Monte Carlo results have become available for the static limit. In contrast,
very little is known about the frequency dependence, even at long–wavelength.
The developments of TD–CDFT outlined above brought renewed interest to
this long–standing problem.

The first Chapter presents the hydrodynamical form of TD–CDFT in the
long–wavelength limit, and including extension to the nonlinear regime. We
then address the frequency–dependence of the longitudinal and transverse
exchange–correlation kernels, starting from the asymptotic values. The low–
frequency limits of the real parts are related to elastic moduli, which in turn
can be expressed in terms of the parameters of Landau theory of Fermi liquids.
At variance with common belief, the small–frequency long–wavelength limit of
the longitudinal exchange–correlation kernel turns out to be discontinuous: the
limiting value is different if the frequency or the wavevector is set to 0 first.
The imaginary parts are in turn related to generalized viscosity coefficients.
Chapter 2 presents an approximate decoupling of the equation of motion for
the current–current response function, leading to a mode–coupling structure
for the exchange–correlation contributions to the long–wavelength spectrum.
This formal result is then used to study the frequency dependence of the local
field correction at long wavelength. The qualitative behaviour of our micro-
scopically derived curves shows evidence for a strong plasmon–plasmon resonant process and is significantly different from previous estimates, including
perturbative results and the quasi–monotonic interpolation proposed by Gross
and Kohn. We then formulate a simple self–consistent model for the long–
wavelength plasmon dispersion in the homogeneous electron gas which emphasizes the interplay between single–plasmon and two–plasmon excitations,
giving results in better qualitative agreement with experiment than previous
homogeneous electron gas theories.

Static properties of quantum fluids can be precisely determined by Quantum Monte Carlo (QMC) simulations. The ground state energies are computed
almost exactly, and accurate interpolations of the QMC data constitute the
key ingredient of the widely used Local Density Approximation (LDA) to static
DFT. QMC simulations established that both the three dimensional (3D) and
the two dimensional (2D) electron gas exhibit a stable spin–polarized phase
at intermediate densities and gave the first reliable estimate of the crystallization density. QMC gives very precise results also on the structure factor,
the static response function and the momentum distribution, which allow not
only to gain quantitative insight into some many–body problems, but also to
quantitatively assess the reliability of approximate many–body theories by direct comparison. Whereas the three dimensional electron gas has already been
studied by QMC in considerable detail, a number of related problems remain
uncovered. Some of them are addressed in Part II of this thesis, and additional
original Monte Carlo results are used in Chapter 6 to study the phase diagram

of the bilayered 2D electron gas. The Diffusion Monte Carlo (DMC) algorithm adopted here is briefly summarized in Chapter 3.

The charged boson fluid was first studied as a model for superconductivity, and presently finds concrete application in the description of the equation of state of white dwarf star cores and in the study of the fusion of three α–particles in a dense He plasma. It has often been used as a development tool for many–body theories, for example in the early days of QMC. We present an extensive study of the three dimensional charged Bose gas at $T = 0$, including momentum distribution, static response properties and the long–wavelength two–body momentum distribution, related to the single particle excitation energies (Chapter 4). The DMC results are fitted with simple analytic expressions to facilitate their use and compared with existing theoretical results. An analogous work in the two–dimensional case both for bosons and fermions is presented in Chapter 5. While there are no previous QMC results on 2D bosons, a thorough discussion of the phase diagram of 2D fermions was recently given by Rapisarda and Senatore. We extend their work to higher densities and give the first DMC results on the momentum distribution.

Electrons confined in two–dimensional layers received a lot of attention after it became possible to obtain very clean experimental realizations. Such systems are particularly attractive since band structure effects are weak and well controlled, correlations are stronger than in 3D, and both density and geometry can be tuned via the design of the experimental sample. The possibility to grow multi–layered structures gives additional degrees of freedom. Two examples of such systems are addressed in Part III. Chapter 6 discusses the electron–electron bilayer, which has been predicted to undergo a number of intriguing phase transitions. A mix of theoretical considerations and Monte Carlo simulations rules out some of this exotic phases and gives accurate predictions for the $T = 0$ phase diagram. We also discuss some experimental results whose interpretation was unclear and find good agreement with a simple model based on the Monte Carlo equation of state for 2D electrons.

Application of a strong magnetic field changes completely the physics of 2D electrons, giving rise to the exciting Quantum Hall (QH) effects. The magnetic field quenches the kinetic energy and effectively lowers the dimensionality of the system, which both result in enhanced correlations. Tunneling experiments provide a natural way to probe the consequences on the electronic spectral function, which describes the single electron excitation spectrum. Whereas vertical tunneling into a homogeneous QH fluid is rather well understood, tunneling into the edges is less clear. A non–Fermi–liquid behaviour was predicted by Wen for Fractional Quantum Hall "sharp" edges, where the density changes from the bulk value to zero along a distance not larger than the magnetic length and only one branch of collective edge excitations is present. Electrostatically confined samples have a quite different "smooth" density profile, and in this case classical hydrodynamics predicts a number of branches of edge collective excitations, some of which have been experimentally observed. In Chapter 7 we address the problem of tunneling into smooth edges, within an Independent Boson Model which emphasizes the interplay of single–particle and collective excitations and predicts significant changes in the shape of the tunneling density of states with varying edge width.

Part I

Long wavelength dynamics in 3D quantum plasmas

Chapter 1

Introduction and rigorous properties

1.1 Introduction

Density Functional Theory (DFT) is to date one of the most successful tools to study static properties of inhomogeneous many–electron systems in which exchange and correlation effects are nonnegligible [1–4]. The widely used Local Density Approximation (LDA) allows to express the unknown exchange and correlation (xc) potential for an inhomogeneous system in terms of the equation of state of the homogeneous electron gas, which is precisely known from Quantum Monte Carlo simulations. The LDA is valid for static systems with slowly varying density, and gives rather good results also in strongly inhomogeneous systems such as atoms and molecules. Quantities which are routinely computed by DFT–LDA include ground state energy and density, equilibrium geometry and ionization energies.

Paralleling a widespread use of this well–established technique, DFT has recently been extended to time–dependent densities (Time Dependent DFT) [5,6] and therefore allows in principle the study of dynamical properties. A number of attempts has been made to formulate a time–dependent extension of the LDA. At sufficiently low frequencies the so–called Adiabatic Local Density Approximation (ALDA) [7–10] has given very useful results. The search for a local dynamical approximation to the xc potential [11–13] has been frustrated by the appearance of various inconsistencies [6,14–16], which can be tracked down to the non–existence of a gradient expansion for the frequency–dependent xc potential in terms of the density alone [17].

Recently Vignale and Kohn (VK) have shown how a local approximation for the xc vector potential can be obtained within Current Density Functional Theory [18,19]. Their expression includes time–dependent xc effects through the long–wavelength xc kernels $f_{xc}^{L,T}(\omega)$ of the homogeneous electron gas, is consistent with the basic conservation laws by construction and becomes exact in the linearized long–wavelength limit. In Section 1.2 we briefly discuss their result, which is interpreted in terms of generalized viscosity coefficients. Standard hydrodynamical arguments allow to uniquely determine the nonlinear generalization up to second order in the spatial gradients, which also turns out to be expressed in terms of the xc kernels $f_{xc}^{L,T}$.

The VK result brought new interest to the frequency dependence of the xc kernels $f_{xc}^{L,T}(\omega)$ of the uniform electron gas. An interpolation between the known asymptotic behaviours had been proposed for the longitudinal component at long wavelength by Gross and Kohn [11–13] and later extended to finite wavevectors by Dabrowski [20]. For the transverse component, a first–order perturbative estimate and the long–wavelength infinite–frequency limit were obtained by Vignale and Kohn [18,19]. This Chapter contains a discussion of the general properties of the kernels $f_{xc}^{L,T}$, including a review of the literature and new results on the high frequency limit at arbitrary wavevector (Section 1.4) and on the long–wavelength low–frequency limit for both components (Section 1.5). Chapter 2 gives a microscopic evaluation of the frequency dependence of both components based on an approximate treatment of multi pair excitations, with results significantly different from both perturbative results and the interpolation of Gross and Kohn for the longitudinal component.

The dynamical exchange and correlation kernels are directly connected with measurable quantities in simple metals, where Electron Energy Loss and Inelastic X–ray scattering experiments allow to extract the plasmon dispersion relation [21,22] and the full dynamic structure factor [23]. However, lattice effects are non–negligible in all metals, and the ones with stronger correlation effects (i.e. Rubidium and Caesium) are particularly difficulty to study experimentally. The measured values of the plasmon dispersion coefficient exhibit a strong dependence on density, which existing theoretical literature is unable to explain as an effect of correlation alone (see Section 2.4.1). The self–consistent model presented in Section 2.4 gives a qualitatively better agreement, still being unable to describe quantitatively the experimental data. A full–fledged TD–DFT computation including band structure effects and the VK xc potential is still to be done.

This Chapter is organized as follows. We first discuss the recent developments in the formulation of a local dynamical approximation to DFT (Section 1.2), which constitute the major motivation for interest in the xc kernels $f_{xc}^{L,T}$. The response–function approach is presented in Section 1.3, both for the density–density and the current–current response and leads to an independent, although equivalent, definition of $f_{xc}^{L,T}$ (Section 1.4). The chapter is concluded by a study of the low–frequency and long–wavelength behaviour of these kernels (Section 1.5), which gives an exact expression of $f_{xc}^{L,T}(k=0,\omega)$ at small ω in terms of the parameters entering Landau theory of Fermi liquids.

1.2   Time–Dependent Density Functional Theory

TD–DFT is based on an exact one to one mapping of the interacting problem into an analogous noninteracting problem with the same density $\rho(\mathbf{r}, t)$ in a fictitious external potential $V^{\text{eff}}[\rho](\mathbf{r}, t)$. The effective potential V^{eff} is a functional of the exact density $\rho(\mathbf{r}, t)^*$ usually written as

$$V^{\text{eff}}[\rho](\mathbf{r}, t) = V^{\text{ext}}(\mathbf{r}, t) + V^{\text{Hartree}}[\rho](\mathbf{r}, t) + V^{xc}[\rho](\mathbf{r}, t) \qquad (1.1)$$

where V^{ext} is the external potential and $V^{\text{Hartree}}[\rho]$ the Hartree self–consistent field. The major theoretical problem is the derivation of reliable approximations for the unknown exchange–correlation contribution $V^{xc}[\rho](\mathbf{r}, t)$. Discussion has mostly focussed on the long–wavelength limit because it allows to use what is known on the homogeneous system and gives a somewhat controlled expansion in terms of the hopefully small parameter $\nabla\rho/\rho$. In the static case this approach (LDA) gives reliable results even if the expansion parameter $\nabla\rho/\rho$ is not small.

If the external potential is slowly dependent on time one can resort to a direct generalization of the static LDA,

$$V^{xc}_{ALDA}[\rho](r, t) = \left. \frac{d[n\varepsilon^{\text{hom}}_{xc}(n)]}{dn} \right|_{n=\rho(\mathbf{r},t)} \qquad (1.2)$$

known as Adiabatic Local Density Approximation (ALDA) [7–10]. Here $\varepsilon^{\text{hom}}_{xc}(n)$ is the xc contribution to the ground state energy per particle of a homogeneous electron gas with density n. We are interested in the dynamical corrections to ALDA.

A non–adiabatic local approximation for the dynamic xc potential in the linearized regime has been proposed by Gross and Kohn (GK) [11]. For a weak long–wavelength density modulation $\rho^{\text{ind}}(\mathbf{r}, t)$ they argue that

$$V^{xc}_{GK}[\rho](\mathbf{r}, \omega) \simeq \rho^{\text{ind}}(\mathbf{r}, t) f^{L}_{xc}(k = 0, \omega, n = \rho(\mathbf{r}, t)) \qquad (1.3)$$

where $f^{L}_{xc}(k, \omega, n)$ is the exchange–correlation kernel of a homogeneous electron gas with density n, defined in Section 1.4. Dynamical effects are included through the frequency–dependence of f^{L}_{xc}, for which GK proposed an interpolation between the known asymptotic behaviours. Their interpolation is discussed in Section 2.3, here we focus on the validity of equation (1.3) itself.

---

*The theorem states that it is a functional of the exact time-dependent density *and* of the *initial state* Ψ. Here and below we follow common practice and leave the dependence on Ψ implicit. Note that if the system starts from the ground state, $\Psi = \Psi_{GS}$ itself is a functional of the density at time $t = 0$, by the Hohenberg-Kohn theorem [1], and therefore $V^{\text{eff}}(\mathbf{r}, t)$ depends only on $\rho(\mathbf{r}, t)$.

During the last decade it has been shown by Dobson [14,15] and Vignale [16,17] that the GK expression (1.3) fails to satisfy a number of general constraints, essentially deriving from Newton's third law. For example equation (1.3) predicts a coupling between the xc force and the center of mass degree of freedom in a system exhibiting rigid oscillatory motion in an external harmonic potential, and therefore violates a generalization of Kohn's theorem known as Harmonic Potential Theorem [14,16]. In 1995 Vignale [17] has shown that the time–dependent V^{xc} does not admit a gradient expansion in terms of the density, thereby proving that further search for a dynamical approximation to $V^{xc}[\rho](\mathbf{r}, t)$ expressed in terms of the local $\rho(\mathbf{r}, t)$ was hopeless.

A fully consistent local approximation has been devised in 1996 by Vignale and Kohn (VK) [18,19]. They propose to use the current density instead of the density as an independent variable. On the one hand, this makes the theory more cumbersome, since it requires an exchange–correlation vector potential $\mathbf{A}^{xc}(\mathbf{r}, \omega)$ which includes a longitudinal (L) and two transverse (T) components. On the other hand, it allows to obtain an expression for the xc potential which is local and consistent by construction with Newton's laws and derived relations. In the linearized long–wavelength limit they obtain the exact result

$$
\begin{aligned}
\mathbf{A}_i^{xc}(\mathbf{r}, \omega) \;=\; & -\frac{1}{\omega^2} \Big\{ \partial_i [f_{xc}^L \nabla \cdot (n_0 \mathbf{u}) - \delta f_{xc}^L \mathbf{u} \cdot \nabla n_0] + \delta f_{xc}^L (\partial_i n_0) \nabla \cdot \mathbf{u} \\
& + f_{xc}^T [-n_0 (\nabla \times \nabla \times \mathbf{u})_i + 2(\partial_j u_i + \partial_i u_j)\partial_j n_0 - 4(\partial_i n_0)\nabla \cdot \mathbf{u}] \\
& + n_0 \left((\partial_j f_{xc}^T)(\partial_i u_j + \partial_j u_i) - 2(\partial_i f_{xc}^T)\nabla \cdot \mathbf{u} \right) \Big\} \;.
\end{aligned} \tag{1.4}
$$

Here, $n_0(\mathbf{r})$ is the unperturbed ground–state density, $\mathbf{u}(\mathbf{r}, \omega) = \mathbf{j}_1(\mathbf{r}, \omega)/n_0(\mathbf{r})$ is the local velocity and $\delta f_{xc}^L(\omega, n) = f_{xc}^L(\omega, n) - f_{xc}^L(0, n)$. The xc kernels of the homogeneous electron gas $f_{xc}^{L,T}$ are functions of ω and of the local density $n_0(\mathbf{r})$ defined in Section 1.4.

Vignale and Kohn derived equation (1.4) above as the unique expression which is linear in $\mathbf{u}$, second-order in the gradients, and consistent with translational and rotational invariance as well as Newton's third law, which states that the total xc force and torque exerted by a volume element on itself vanish. These are the same constraints used in the standard derivation of the hydrodynamical equations (see, e.g. Ref. [24]), which we now summarize.

Local conservation of momentum (the first part of Newton's third law) implies that the total xc force acting on a given volume element can be written as the surface integral of a local quantity, which is defined to be the flux of the stress tensor $\sigma_{ij}(\mathbf{r}, t)$. Equivalently, force density is the divergence of σ_{ij}. Conservation of angular momentum (the second part of Newton's third law) then implies that $\sigma_{ij} = \sigma_{ji}$: the total torque exerted by a volume element to itself must vanish.[†] The stress tensor $\sigma_{ij}(\mathbf{r}, t)$ can depend only on local

---

[†]The torque exerted on a volume element V is $\int_V \epsilon_{ijk} \mathbf{r}_j \nabla_l \sigma_{lk}$. Integration by parts shows that this is composed by a surface contribution and by the internal contribution $- \int_V \epsilon_{ijk} \sigma_{jk}$, which must vanish.

density and velocity. Invariance under Galilean transformations implies[‡] that the velocity can enter σ_{ij} only through its gradient, which is small at long wavelength even if the velocity itself is not small. In the linear regime one concludes that the xc force acting on a particle $\mathbf{F}_{xcl} = -i\omega\mathbf{A}_{xcl}$ is given by

$$\mathbf{F}_{xcl}(\mathbf{r},\omega) = -\nabla_i V_{xcl}^{ALDA}(\mathbf{r},\omega) + \frac{1}{n_0(\mathbf{r})}\sum_j \frac{\partial\sigma_{xc,ij}(\mathbf{r},\omega)}{\partial\mathbf{r}_j} \qquad (1.5)$$

where the first term is the linearization of the ALDA expression (1.2), and the dynamical correction is the divergence of the visco-elastic stress tensor

$$\sigma_{xc,ij} = \tilde{\eta}_{xc}\left(\frac{\partial\mathbf{u}_i}{\partial\mathbf{r}_j} + \frac{\partial\mathbf{u}_j}{\partial\mathbf{r}_i} - \frac{2}{3}\nabla\cdot\mathbf{u}\delta_{ij}\right) + \tilde{\zeta}_{xc}\delta_{ij}\nabla\cdot\mathbf{u}\,. \qquad (1.6)$$

Here $\tilde{\eta}_{xc}(\omega,n_0(\mathbf{r}))$ and $\tilde{\zeta}_{xc}(\omega,n_0(\mathbf{r}))$ are complex viscosity coefficients, and equation (1.6) has been obtained as the most general second-order symmetric tensor which depends linearly on $\partial_i\mathbf{u}_j$. Straightforward computations show that this relation is identical to the Vignale–Kohn result (1.4) if

$$\tilde{\zeta}_{xc}(\omega,n) = -\frac{n^2}{i\omega}\left[f_{xc}^L(\omega,n) - \frac{4}{3}f_{xc}^T(\omega,n) - \frac{d^2\epsilon_{xc}(n)}{dn^2}\right] \qquad (1.7)$$

and

$$\tilde{\eta}_{xc}(\omega,n) = -\frac{n^2}{i\omega}f_{xc}^T(\omega,n)\,. \qquad (1.8)$$

Elasticity theory [25] connects the imaginary parts of $\tilde{\eta}$ and $\tilde{\zeta}$ to the bulk and shear moduli of an isotropic elastic medium, via $K_{xc}^{dyn}(\omega) = \omega\text{Im}\tilde{\zeta}_{xc}$ and $\mu_{xc}^{dyn}(\omega) = \omega\text{Im}\tilde{\eta}_{xc}$[§] and therefore connects directly equations (1.7-1.8) with the espressions given in Section 1.5.3 below. For $\omega \to 0$ K_{xc}^{dyn} vanishes, while μ_{xc}^{dyn} has a finite value (see Section 1.5.3). A similar state of affairs holds for the noninteracting *kinetic* contributions to the bulk and shear moduli: $K_{kin}^{dyn} = 0$ and $\mu_{kin}^{dyn} = p(n)$, where $p(n)$ is the noninteracting Fermi pressure, and $K_{kin}^{stat} = ndp(n)/dn$, $\mu_{kin}^{stat} = 0$. The general conclusion is that dynamical (post-ALDA) effects do not modify the bulk modulus, but they cause the appearance of a nonvanishing shear modulus and viscosity.

The above discussion can be generalized to the nonlinear regime. The xc force, related to the xc vector potential $\mathbf{A}_{xc}$ by

$$\mathbf{F}_{xc,i} = n\left[\left(\frac{\partial}{\partial t} + \mathbf{u}\cdot\nabla\right)\mathbf{A}_{xc,i} - \sum_j \mathbf{u}_j\nabla_i\mathbf{A}_{xc,j}\right], \qquad (1.9)$$

[‡] Indeed, the *forces* should not change in going to a frame moving with a constant velocity - therefore adding a constant to $\mathbf{u}$ should not change σ_{ij}.

[§] The superscript *dyn* is a reminder that these are *dynamical* contributions to be added to the usual static ones, $K_{xc}^{stat} = n^2 d^2\epsilon_{xc}(n)/dn^2$ and $\mu_{xc}^{stat} = 0$, already present in the ALDA.

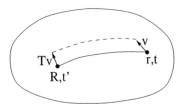

FIGURE 1.1: The fluid element at position $\mathbf{R}(t'|\mathbf{r},t)$ at time t' evolves into the one at $\mathbf{r}$ at time t. The fluid element at position $\mathbf{R}+T\mathbf{v}$ at time t' evolves into the one at position $\mathbf{r}+\mathbf{v}$ at time t (see text).

must be the divergence of a local symmetric rank two stress tensor. The concept of "locality" is ambiguous for retarded phenonema, and only a definition independent on the reference frame¶ guarantees consistency by construction with the transformation laws under translational [26] and rotational accelerations. Let $\mathbf{R}(t'|\mathbf{r},t)$ be the position at time t' of the fluid element which evolves into $\mathbf{r}$ at time t, and $T(t'|\mathbf{r},t)\mathbf{v}$ be the vector embedded in the fluid at time t' and position $\mathbf{R}(t'|\mathbf{r},t)$ which evolves into $\mathbf{v}$ at position $\mathbf{r}$ at time t (see Figure 1.1)‖. Then, $\sigma_{ij}(\mathbf{r},t)$ depends on $\rho(\mathbf{R},t')$, $T(t|\mathbf{R},t')\mathbf{j}(\mathbf{R},t')$ and their spatial derivatives. Use of the retarded position and orientation in the equations below allows to satisfy exactly by construction the zero–force and zero–torque theorems, but the full result differs from the naive estimate $\mathbf{R}\simeq\mathbf{r}$ and $T\simeq$ identity only in higher order terms in the gradient expansion. For ease of notation we adopt this simpler prescription in the following equations.

The above requirements, and the fact that the familiar Navier–Stokes equation must be recovered for slowly varying (in time) perturbations, uniquely determine the form of $\mathbf{A}_{xc}$ to second order in the spatial derivatives:

$$\frac{\partial \mathbf{A}_{xc,i}(\mathbf{r},t)}{\partial t} = -\nabla_i v_{xc}^{ALDA}(\mathbf{r},t)+\frac{1}{\rho(\mathbf{r},t)}\sum_j \frac{\partial \sigma_{xc,ij}(\mathbf{r},t)}{\partial r_j}, \qquad (1.10)$$

where

$$\sigma_{ij}(\mathbf{r},t) = \int_{-\infty}^{t}\left[\tilde{\eta}(\rho(\mathbf{r},t),t-t')\left(\frac{\partial u_i(\mathbf{r},t')}{\partial r_j}+\frac{\partial u_j(\mathbf{r},t')}{\partial r_i}-\frac{2}{3}\nabla\cdot\mathbf{u}(\mathbf{r},t')\delta_{ij}\right)\right.$$
$$\left.+\tilde{\zeta}(\rho(\mathbf{r},t),t-t')\nabla\cdot\mathbf{u}(\mathbf{r},t')\delta_{ij}\right]dt', \qquad (1.11)$$

$\tilde{\eta}(n,t-t') \equiv \int \tilde{\eta}(n,\omega)\exp(-i\omega(t-t'))d\omega/2\pi$, and similarly for $\tilde{\zeta}$. Here $\rho(\mathbf{r},t)$ and $\mathbf{u}(\mathbf{r},t)$ are the time-dependent values of the density and velocity field. The

---

¶In the sense that the statement "$\sigma(\mathbf{r},t)$ depends on $\rho(\mathbf{r}',t')$" depends on the motion of the reference frame beween time t and time t'.

‖With obvious notation, for small ϵ one has $\mathbf{R}(t'|\mathbf{r}+\epsilon\mathbf{v},t) - \mathbf{R}(t'|\mathbf{r},t) = T\epsilon\mathbf{v}+O(\epsilon^2)$. Furthermore, the inverse matrix of $T(t'|\mathbf{r},t)$ is $T(t|\mathbf{r}',t')$, where $\mathbf{r}' = \mathbf{R}(t'|\mathbf{r},t)$.

apparent ambiguity of whether the density entering the viscosity coefficients in equation (1.11) should be evaluated at time t or at some earlier time t', is resolved by noting that the difference $\rho(\mathbf{r}, t') - \rho(\mathbf{r}, t) = \int_{t'}^{t} \nabla \cdot \mathbf{j}(\mathbf{r}, \tau) d\tau$ generates a higher order gradient correction, provided that the range of times which contribute significantly to the integral in equation (1.11) is essentially finite.

We finally remark that the xc kernels of the homogeneous electron gas $f_{xc}^{L,T}$ fully determine the dynamic effective potential $\mathbf{A}_{xc}(\mathbf{r}, t)$ of Current DFT in the linear (eqs. 1.5–1.8) *and in the nonlinear* (eqs. 1.10–1.11) regime, provided both density and velocity are slowly varying *in space*. This observation provides the major motivation for the detailed study of $f_{xc}^{L,T}$ presented in the remaining part of this Chapter and in the next one.

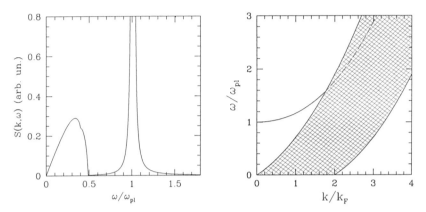

FIGURE 1.2: Left panel: dynamic structure factor at small wavevector $kr_0 = 0.4$ at $r_s = 1$ as a function of ω/ω_{pl}, displaying the particle–hole continuum at low frequency and the strong plasmon resonance at $\omega = \omega_{pl}$, broadened by damping. Right panel: excitation spectrum, neglecting plasmon broadening

1.3   Density and current response functions

This section presents the response–function approach to the study of the homogeneous electron gas, including both the density response and the current response. Some basic results are briefly recalled, with attention to the aspects which will be used to define and study the xc kernels $f_{xc}^{L,T}$.

1.3.1   Density–density response function

The density–density response function $\chi_{\rho\rho}(\mathbf{k}, \omega)$ [27] describes the density response to a weak external potential which couples to the density, i.e. to a scalar potential in the case of a charged system. It is defined as

$$\rho^{\mathrm{ind}}(\mathbf{k}, \omega) = \chi_{\rho\rho}(\mathbf{k}, \omega) V^{\mathrm{ext}}(\mathbf{k}, \omega) \tag{1.12}$$

where V^{ext} and ρ^{ind} are the external potential and the induced density modulation, respectively. Its main properties are briefly summarized below.

Perfect screening: charged fluids screen completely long–wavelength static external potentials. Ideed, $\chi_{\rho\rho}(\mathbf{k} \to 0, \omega = 0) \simeq -k^2/4\pi e^2$. This *static* long–wavelength behaviour is not continuously connected with the finite frequency long–wavelength limit $\chi_{\rho\rho}(\mathbf{k} \to 0, \omega) \simeq nk^2/m(\omega^2 - \omega_{pl}^2)$.

Collective excitations: the poles of the response function $\chi_{\rho\rho}(\mathbf{k}, \omega)$, seen as a function of ω at a fixed $\mathbf{k}$, describe the collective excitations, namely the plasmon for an electron gas at small $\mathbf{k}$. In the long–wavelength limit the plasmon has frequency $\omega_{pl} = \sqrt{4\pi e^2 n/m}$ and is not damped.

Fluctuation–dissipation theorem: The dynamic structure factor $S(k,\omega)$, proportional to $\mathbf{Im}\,\chi_{\rho\rho}(\mathbf{k},\omega)$, describes the spectral strength of density excitations, as measured e.g. by inelastic X-ray scattering experiments. It is known to contain two main features: a sharp plasmon resonance (present only at small k) and a broad single–pair continuum (see Figure 1.2). In Section 2.3 we show that there is an additional structure at small k corresponding to double–plasmon excitations.

Causality: In time domain the density modulation has to follow the potential which determined it, therefore $\chi_{\rho\rho}(\mathbf{k},t) = 0$ for $t < 0$. This leads to the Kramers–Kronig relation:

$$\chi(k,\omega) = \frac{1}{\pi}\mathcal{P}\int_{-\infty}^{\infty}\frac{\mathbf{Im}\,\chi(k,\omega')}{\omega' - \omega - i\eta}d\omega' \qquad (1.13)$$

where η is a positive infinitesimal, which is indeed valid for any response function.

1.3.2   Current–current response function

The above discussion can be straightforwardly generalized to the current response to an applied vector potential $\mathbf{A}^{\text{ext}}(\mathbf{k},\omega)$. The response function is defined in terms of the induced canonical (paramagnetic) current $\mathbf{j}(\mathbf{k},\omega)$,

$$\mathbf{j}_i^{\text{ind}}(\mathbf{k},\omega) = \chi_{ij}(k,\omega)\mathbf{A}_j^{\text{ext}}(\mathbf{k},\omega) \qquad (1.14)$$

where the factor of e/c has been included in the vector potential $\mathbf{A}$ and the gauge $V^{\text{ext}}(\mathbf{k},\omega) = 0$ has been used**. The physical current $\tilde{\mathbf{j}} = \mathbf{j} + (n/m)\mathbf{A}$ is given by the gauge–invariant expression

$$\tilde{\mathbf{j}}_i(\mathbf{k},\omega) = \left(\chi_{ij}(\mathbf{k},\omega) + \frac{n}{m}\delta_{ij}\right)\left[\mathbf{A}_j^{\text{ext}}(\mathbf{k},\omega) - \frac{\mathbf{k}_j}{\omega}V^{\text{ext}}(\mathbf{k},\omega)\right] \qquad (1.15)$$

where n is the number density (see Ref. [27], Section 4.7).

In a homogeneous and isotropic system the tensor response function has only two independent components, usually called longitudinal (χ_L) and transverse (χ_T), defined from

$$\chi_{ij}(\mathbf{k},\omega) = \delta_{ij}\chi_T(k,\omega) + \hat{\mathbf{k}}_i\hat{\mathbf{k}}_j\left(\chi_L(k,\omega) - \chi_T(k,\omega)\right). \qquad (1.16)$$

**A scalar potential $V(\mathbf{k},\omega)$ can always be written as a vector potential $\mathbf{A} = -\mathbf{k}V(\mathbf{k},\omega)/\omega$.

Gauge invariance and the continuity equation allow to relate the density–density response function to the longitudinal component of the current–current response function [27],

$$\chi_{\rho\rho}(k,\omega) = \frac{nk^2}{m\omega^2} + \frac{k^2}{\omega^2}\chi_L(k,\omega).$$ (1.17)

The first term on the left hand side of the above equation is the leading high–frequency behaviour and is equivalent, via Kramers–Kronig, to the well-known f–sum rule.

All that has been said concerning the density–density response function $\chi_{\rho\rho}$ applies through (1.17) to the longitudinal component of χ_{ij}. The only significant difference in the transverse case is the absence of a long–range transverse potential (in the absence of retardation, i.e. for $c \to \infty$). As a consequence, there is no perfect screening condition and no plasmon–like excitation.

The density–density response function $\chi_{\rho\rho}^0$ of non–interacting electrons was computed by Lindhard [28], and the longitudinal current–current χ_L^0 is obtained from equation (1.17). An analogous result for the transverse component χ_T^0 is presented in Appendix B.

1.3.3   Equation of motion for the response functions

The response function describing the first–order variation in the expectation value of an operator A under application of a potential coupled to an operator B, i.e. under a perturbation hamiltonian $H' = V^{\text{ext}}B$, is given by

$$\chi_{AB}(\omega) = \ll A; B \gg_\omega = -i \int_0^\infty dt\, e^{i(\omega+i\eta)t} \langle [A(t), B(0)] \rangle$$ (1.18)

where $A(t) = e^{iHt}Ae^{-iHt}$ (here and below, $\hbar = 1$). Integration by parts gives the equations of motion

$$\ll A; B \gg = \frac{1}{\omega}\langle [A,B] \rangle + \frac{1}{\omega}\ll [A,H]; B \gg_\omega =$$ (1.19)

$$= \frac{1}{\omega}\langle [A,B] \rangle - \frac{1}{\omega}\ll A; [B,H] \gg_\omega .$$ (1.20)

Considering the density and current operators

$$\rho_{\mathbf{k}} = \sum_{\mathbf{q},\sigma} c_{\mathbf{q}-\mathbf{k}/2,\sigma}^\dagger c_{\mathbf{q}+\mathbf{k}/2,\sigma}$$ (1.21)

$$\mathbf{j}_{\mathbf{k}} = \sum_{\mathbf{q},\sigma} \frac{\mathbf{q}}{m} c_{\mathbf{q}-\mathbf{k}/2,\sigma}^\dagger c_{\mathbf{q}+\mathbf{k}/2,\sigma}$$ (1.22)

we have $\chi_{\rho\rho}(\mathbf{k},\omega) = \ll \rho_{\mathbf{k}}; \rho_{-\mathbf{k}} \gg_\omega$ and $\chi_{ij}(\mathbf{k},\omega) = \ll \mathbf{j}_{\mathbf{k}}^{(i)}; \mathbf{j}_{-\mathbf{k}}^{(j)} \gg_\omega$, and equation (1.17) is equivalent to the operator form of the continuity equation in the absence of external fields $[\rho_{\mathbf{k}}, H] = \mathbf{k} \cdot \mathbf{j}_{\mathbf{k}}$.

The full electron gas Hamiltonian is $H = KE + PE$, where

$$KE = \sum_{\mathbf{k},\sigma} \frac{k^2}{2m} c^\dagger_{\mathbf{k},\sigma} c_{\mathbf{k},\sigma} \tag{1.23}$$

and

$$PE = \frac{1}{2V} \sum_{\mathbf{q}\neq 0} v_q (\rho_{-\mathbf{q}}\rho_{\mathbf{q}} - N) . \tag{1.24}$$

Here and below $v_k = 4\pi e^2/k^2$ is the Fourier transform of the Coulomb potential, whose $\mathbf{q} = 0$ term is excluded from equation (1.24) due to the global charge neutrality constraint.

In the equation of motion for $\mathbf{j}_\mathbf{k}$ the potential energy terms with $\mathbf{q} = \pm\mathbf{k}$ are singular and need to be treated separately. For longitudinal currents we get

$$[[\mathbf{k} \cdot \mathbf{j}_\mathbf{k}, PE^{\text{res}}], H] = \omega^2_{pl} \mathbf{k} \cdot \mathbf{j}_\mathbf{k} \tag{1.25}$$

where the resonant term $PE^{\text{res}} = V^{-1} v_k \rho_{-\mathbf{k}}\rho_\mathbf{k}$. This proves that excluding PE^{res} from the equation of motion for $\chi_L(k,\omega)$ we obtain an expression for the proper response function $\tilde{\chi}_L(k,\omega)$ [27,29] defined by

$$\left(\tilde{\chi}_L(k,\omega) + \frac{n}{m}\right)^{-1} = \left(\chi_L(k,\omega) + \frac{n}{m}\right)^{-1} - v_k . \tag{1.26}$$

The transverse component of $\mathbf{j}_\mathbf{k}$ instead commutes with PE^{res}, reflecting the fact that $\tilde{\chi}_T = \chi_T$. In the following we shall deal only with proper terms.

The second time derivative of $\tilde{\chi}_{ij}(k,\omega)$ fixes its high frequency behaviour (see Appendix A):

$$\omega^2 \tilde{\chi}_{ij}(\mathbf{k},\omega) = \tilde{M}^{ij}_3(k) - \ll [j^{(i)}_\mathbf{k}, H]; [j^{(j)}_{-\mathbf{k}}, H] \gg_\omega \tag{1.27}$$

where the last term vanishes for $\omega \to \infty$,

$$\tilde{M}^L_3(k) = \frac{nk^2}{2m^2} \left[\frac{k^2}{2m} + 4\langle ke \rangle + \frac{2}{n} \sum_\mathbf{q} v_q \frac{(\mathbf{k} \cdot \mathbf{q})^2}{k^4} [S(\mathbf{q}+\mathbf{k}) - S(\mathbf{q})] \right] \tag{1.28}$$

and

$$\tilde{M}^T_3(k) = \frac{nk^2}{2m^2} \left[\frac{4}{3}\langle ke \rangle + \frac{1}{n} \sum_\mathbf{q} v_q \left[\frac{q^2}{k^2} - \frac{(\mathbf{k} \cdot \mathbf{q})^2}{k^4} \right] \cdot [S(\mathbf{q}+\mathbf{k}) - S(\mathbf{q})] \right] \tag{1.29}$$

$\langle ke \rangle$ denoting the exact kinetic energy per particle. The high–frequency limit of the full response function is then obtained from $M^L_3 = \tilde{M}^L_3 + \omega^2_{pl} n/m$, $M^T_3 = \tilde{M}^T_3$, and Kramers–Kronig relates these asymptotic behaviours to the spectral moments

$$\tilde{M}^{L,T}_3(k) = -\frac{2}{\pi} \int_0^\infty \omega \mathbf{Im}\, \tilde{\chi}_{L,T}(k,\omega) d\omega . \tag{1.30}$$

The expression (1.28) for $\tilde{M}^L_3$ is identical to the usual result for the third moment of the density-density response function divided by k^2 [30]. The transverse extension (1.29) is new, and in the long–wavelength limit is equivalent to the expression for $f^{xc}_T(k = 0, \omega = \infty)$ obtained in Ref. [18].

1.4 Exchange–Correlation Kernels $f_{xc}^{L,T}(k,\omega)$

The xc kernels $f_{xc}^{L,T}$ are defined as

$$
\frac{k^2}{\omega^2} f_{xc}^{L,T}(k,\omega) = \left(\chi_{L,T}^0(k,\omega) + \frac{n}{m}\right)^{-1} - \left(\chi_{L,T}(k,\omega) + \frac{n}{m}\right)^{-1} - v_k^{L,T} \quad (1.31)
$$

$$
= \left(\chi_{L,T}^0(k,\omega) + \frac{n}{m}\right)^{-1} - \left(\tilde{\chi}_{L,T}(k,\omega) + \frac{n}{m}\right)^{-1} \quad (1.32)
$$

where $v_k^L = v_k = 4\pi e^2/k^2$ and $v_k^T = 0$. If $k \ll \omega$ the proper response functions $\tilde{\chi}_{L,T}$ tend to zero as k^2/ω^2, and to leading order

$$
f_{xc}^{L,T}(k,\omega) = \frac{\omega^2 m^2}{k^2 n^2} \left[\tilde{\chi}_{L,T}(k,\omega) - \chi_{L,T}^0(k,\omega)\right]. \quad (1.33)
$$

The long wavelength limit $f_{xc}(k = 0,\omega)$, which we denote by $f_{xc}(\omega)$, is given by

$$
\mathbf{Im}\, f_{xc}^{L,T}(\omega) = \lim_{k \to 0} \frac{\omega^2 m^2}{k^2 n^2} \mathbf{Im}\, \tilde{\chi}_{L,T}(k,\omega). \quad (1.34)
$$

We further note that $f_{xc}^{L,T}$ are causal functions which obey Kramers–Kronig relations (see equation (2.5)).

The exchange–correlation contribution to the density–density response function is often expressed in terms of the so-called local field factor $G(k,\omega) = -f_{xc}^L(k,\omega)/v_k^L$. The traditional definition

$$
G(k,\omega) = \frac{1}{v_k \chi_{\rho\rho}(k,\omega)} - \frac{1}{v_k \chi_{\rho\rho}^0(k,\omega)} - 1 \quad (1.35)
$$

is equivalent to the longitudinal part of (1.31) through (1.17). The Random Phase Approximation (RPA) corresponds to $G(k,\omega) = 0$, and a vast class of approximations, known as static local field approximations, neglect the frequency dependence of $G(k,\omega)$ and adopt various approximate expressions for $G(k)$. The $\omega = 0$ limit $G(k,\omega = 0)$ describes the static response, and will be discussed in Section 4.2 in the case of the bosonic plasma. Here we are mainly interested in the behaviour of f_{xc} as a function of ω at $k = 0$.

Equations (1.28) and (1.29) fix the high-ω limit

$$
f_{xc}^{L,T}(k,\omega = \infty) = \frac{m^2}{k^2 n^2} \tilde{M}_3^{L,T}(k) \quad (1.36)
$$

which at long wavelength gives (see also equation (A.4))

$$
f_{xc}^L(\omega = \infty) = \frac{1}{2n}\left[4\left(\langle ke\rangle - \langle ke\rangle^0\right) + \frac{8}{15}\langle pe\rangle\right] \quad (1.37)
$$

$$
f_{xc}^T(\omega = \infty) = \frac{1}{2n}\left[\frac{4}{3}\left(\langle ke\rangle - \langle ke\rangle^0\right) - \frac{4}{15}\langle pe\rangle\right] \quad (1.38)
$$

where $\langle ke \rangle$ and $\langle pe \rangle$ denote the exact kinetic and potential energies, and $\langle ke \rangle^0$ the corresponding ideal–gas quantity.

The first order perturbative result for $f_{xc}^{L,T}(\omega)$ can be immediately obtained from equations (1.37-1.38) above. In fact, to first order $\mathbf{Im}\, f_{xc}^{L,T}(\omega)$ vanishes identically for $\omega \neq 0$ due to phase space constraints, therefore by Kramers–Kronig the first–order $\mathbf{Re}\, f_{xc}^{L,T}(\omega)$ is constant and equals the high–frequency limit. Hartree–Fock gives $\langle ke \rangle^1 = \langle ke \rangle^0$ and $\langle pe \rangle^1 = -3(18\pi)^{1/3} Ry/4\pi r_s$ (from exchange), and therefore to first order $f_{xc}^L(\omega) = -3\pi e^2/5 k_F^2$ and $f_{xc}^T(\omega) = 3\pi e^2/10 k_F^2$, for $\omega \neq 0$. More interesting results are obtained by second–order perturbation theory, and will be briefly discussed in Chapter 2.

The static long–wavelength limit is fixed by the compressibility sum rule

$$\lim_{k \to 0} f_{xc}^L(k, \omega = 0) = \frac{1}{n^2}\left(\frac{1}{K_T} - \frac{1}{K_T^0}\right) \tag{1.39}$$

where K_T and K_T^0 are the exact and the ideal–gas compressibilities. In the transverse case, the analogous long wavelength static limit gives the orbital diamagnetic susceptibility $\xi = \lim_{\mathbf{k} \to 0} M(\mathbf{k})/H(\mathbf{k})$, where $M(\mathbf{k})$ is the orbital magnetization and $H(\mathbf{k})$ the applied magnetic field. We get

$$\xi = -\frac{e^2}{c^2}\lim_{k \to 0}\frac{\chi_T(k, 0) + n/m}{k^2} = -\frac{e^2}{c^2}\left\{\frac{12m\pi^2}{k_F} - \lim_{k \to 0}\left[\lim_{\omega \to 0}\frac{k^4 f_{xc}^T(k, \omega)}{\omega^2}\right]\right\}^{-1}. \tag{1.40}$$

A perturbative expression for ξ at small r_s has been obtained by Vignale *et al* [31]. The most relevant consequence of equation (1.40) on the present discussion is that at sufficiently small k one has $f_{xc}^T(k, \omega = 0) = 0$.

Table 1.1 reports the numerical evaluation of the exact asymptotic behaviours evaluated from the DMC equation of state from Refs. [32,33]. Note that (1.40) is excluded since no precise simulation results exist on the orbital magnetic susceptibility.

1.5   Low–frequency long–wavelength limit of the xc kernels $f_{xc}^{L,T}$

This section is dedicated to the study of the limit of $f_{xc}^{L,T}(k = 0, \omega)$ for $\omega \to 0$, which turns out not to be what could be naively guessed from equations (1.39) and (1.40). We first compute the long–wavelength and low–frequency behaviour of the current–current response function by means of the Landau theory of Fermi liquids [27], which allows to express the various limits of the xc kernels $f_{xc}^{L,T}$ in terms of the Landau quasiparticle interaction parameters F_l. The connection with elastic constants is made more explicit through the discussion of the response to affine perturbations, which are defined as the perturbations which generate a uniform elastic strain (Sections and 1.5.2 and

TABLE 1.1: Exact limiting behaviours of the longitudinal and transverse $f_{xc}^{L,T}(k = 0, \omega)$ from DMC equation of state [32,33], in units of $2\omega_{pl}/n$.

r_s	$f_{xc}^L(\omega = 0)$	$f_{xc}^L(\omega = \infty)$	$f_{xc}^T(\omega = \infty)$
0.1	-0.01867	-0.00998	0.00640
0.5	-0.04246	-0.01794	0.01771
1	-0.0611	-0.0216	0.0284
2	-0.0891	-0.0252	0.0457
3	-0.1119	-0.0280	0.0600
4	-0.1320	-0.0308	0.0724
5	-0.1503	-0.0338	0.0835
6	-0.1674	-0.0370	0.0935
10	-0.2276	-0.0518	0.1267
12	-0.2544	-0.0599	0.1404
15	-0.2917	-0.0725	0.1587
20	-0.3483	-0.0939	0.1847

1.5.3). The same expressions can be obtained within static Landau theory, i.e. using the expansion of the internal energy in terms of quasiparticle occupation numbers (Section 1.5.4). The results are discussed and evaluated numerically in Section 1.5.5. Apart from Section 1.5.1, all the discussion is made in parallel for two and three spatial dimensions. For ease of notation, we define the "static" and "dynamic" $(k, \omega) \to (0,0)$ limits as

$$f_{xc}^{L,T\,\mathrm{stat}} = \lim_{k\to 0}\lim_{\omega\to 0} f_{xc}^{L,T}(k, \omega), \qquad (1.41)$$

$$f_{xc}^{L,T\,\mathrm{dyn}} = \lim_{\omega\to 0}\lim_{k\to 0} f_{xc}^{L,T}(k, \omega). \qquad (1.42)$$

The main result of this section is $f_{xc}^{L\,\mathrm{dyn}} - f_{xc}^{L\,\mathrm{stat}} = \xi f_{xc}^{T\,\mathrm{dyn}}$ (here, $\xi = 4/3$ in 3D and $\xi = 1$ in 2D), which implies $f_{xc}^{L\,\mathrm{stat}} \neq f_{xc}^{L\,\mathrm{dyn}}$.

1.5.1   Equation of motion for $\delta n_{\mathbf{p}}$ within Landau theory

The equation of motion for the quasiparticle distribution function $n_{\mathbf{p}}$ is (for a derivation see e.g. the book by Pines and Nozières [27])

$$(\mathbf{k} \cdot \mathbf{v_p} - \omega)\delta n_{\mathbf{p}} - \mathbf{k} \cdot \mathbf{v_p}\frac{\partial n_0}{\partial \varepsilon_p}\sum_{\mathbf{p'}} f_{\mathbf{pp'}}\delta n_{\mathbf{p'}} - \mathbf{k} \cdot \mathbf{v_p}\frac{\partial n_0}{\partial \varepsilon_p}\frac{\mathbf{p} \cdot \mathbf{A}^{\mathrm{ext}}}{m} = 0 \qquad (1.43)$$

where $\mathbf{A}^{\mathrm{ext}}(\mathbf{k}, \omega)$ is the external vector potential, $n_0(\varepsilon_{\mathbf{p}})$ the equilibrium quasiparticle momentum distribution, and $\delta n_{\mathbf{p}}(\mathbf{k}, \omega)$ the induced modulation, which gives the current $\mathbf{j}(\mathbf{k}, \omega) = \sum_{\mathbf{p}} \mathbf{p} n_{\mathbf{p}}/m$. Equation (1.43) can be solved with respect ot $\delta n_{\mathbf{p}}$ to give $\delta n_{\mathbf{p}} = \Pi_{\mathbf{p}}\mathbf{p} \cdot \mathbf{A}^{\mathrm{ext}}/m$, where the single-channel response function $\Pi_{\mathbf{p}}$ obeys

$$\Pi_{\mathbf{p}} = R_{\mathbf{p}} + R_{\mathbf{p}} \sum_{\mathbf{p}'} f_{\mathbf{p}\mathbf{p}'} \Pi_{\mathbf{p}'} \qquad (1.44)$$

and

$$R_{\mathbf{p}} = \frac{\partial n_0}{\partial \varepsilon_p} \frac{\mathbf{k} \cdot \mathbf{p}}{\mathbf{k} \cdot \mathbf{p} - m^* \omega} \qquad (1.45)$$

gives the response of noninteracting quasiparticles.

For $k \ll m^* \omega / k_F$ expansion of the denominator in equation (1.45) in powers of $\mathbf{k} \cdot \mathbf{p}/m^* \omega$ shows that up to order k^2 the angular dependence in equation (1.44) is limited to small positive powers of $\mathbf{k} \cdot \mathbf{p}$. Careful separation of the various angular momentum channels and straightforward calculations allow to determine the longitudinal and transverse components of the current–current response functions, which then give

$$f_{xc}^{L\,\mathrm{dyn}} = \frac{2\varepsilon_F}{n}\frac{3}{5}\frac{\frac{5}{9}F_0 + \frac{4}{45}F_2 - \frac{1}{3}F_1}{1 + F_1/3}, \qquad (1.46)$$

$$f_{xc}^{T\,\mathrm{dyn}} = \frac{2\varepsilon_F}{n}\frac{1}{5}\frac{F_2/5 - F_1/3}{1 + F_1/3} \qquad (1.47)$$

where $\varepsilon_F = k_F^2/2m$ is the bare Fermi energy and the angular components F_l of the Landau parameters are defined as usual by

$$f_{\mathbf{p}\mathbf{p}'} = \frac{\pi^2}{m^* k_F} \sum_l P_l(\cos\theta_{\mathbf{p}\mathbf{p}'}) F_l . \qquad (1.48)$$

Application of the same method to the static ($\omega = 0$) case leads to the well–known compressibility sum rule (1.39), which in the language of Landau theory states

$$f_{xc}^{L\,\mathrm{stat}} = \frac{2\varepsilon_F}{n}\frac{1}{3}\frac{F_0 - F_1/3}{1 + F_1/3} . \qquad (1.49)$$

We now proceed to give an alternative derivation, which enlightens the relation between $f_{xc}^{L,T\,\mathrm{dyn}}$ and elasticity theory. The present results are discussed and evaluated numerically in Section 1.5.5 below.

1.5.2 Connection with the canonical response to affine perturbations

In this Subsection we illustrate the connection between the long-wavelength xc kernels $f_{xc}^{L,T}$ and the canonical response to affine perturbations. By canonical response we mean a collisionless response, which can be described as a canonical evolution of the initial state. This is appropriate in the *dynamical* regime. By affine perturbation we mean a transformation of the form

$$\mathbf{r}_i \rightarrow \mathbf{r}_i + \lambda(t) a_{ij} \mathbf{r}_j , \qquad (1.50)$$

where $\lambda(t)$ is a small and slowly varying function of time, a_{ij} is $d \times d$ real matrix (d is the number of spatial dimensions), and sum over repeated cartesian indices is implied. For example, a purely shear deformation would be a translation of the y coordinate of each particle by an amount proportional to the x coordinate, $y \to y + \lambda(t)x$, other coordinates being unaffected.

Under the assumption that relaxation does not occur on the time scale of the transformation, the evolution of the quantum state $|\psi(t)\rangle$ of the system can be viewed as the unfolding of a unitary transformation

$$|\psi(t)\rangle = \exp\left[i\lambda(t)\hat{A}\right]|\psi(0)\rangle, \tag{1.51}$$

where

$$\hat{A} = \sum_{\alpha} a_{ij}\mathbf{p}_i^{\alpha}\mathbf{r}_j^{\alpha} \tag{1.52}$$

is the generator of the deformation, and α is a particle index. Indeed, $[i\lambda\hat{A}, \mathbf{r}_i^{\alpha}] = \lambda a_{ij}\mathbf{r}_j^{\alpha}$.

The connection with the current response is best understood considering the long–wavelength expansion of the current operator in first–quantization form,

$$\mathbf{j}_i = \frac{1}{2m}\sum_{\alpha}\left(\mathbf{p}_i^{\alpha}e^{i\mathbf{k}\cdot\mathbf{r}^{\alpha}} + e^{i\mathbf{k}\cdot\mathbf{r}^{\alpha}}\mathbf{p}_i^{\alpha}\right) \simeq \sum_{\alpha}\frac{\mathbf{p}_i^{\alpha}}{m} + \frac{i\mathbf{k}_j}{2m}\sum_{\alpha}\left(\mathbf{p}_i^{\alpha}\mathbf{r}_j^{\alpha} + \mathbf{r}_j^{\alpha}\mathbf{p}_i^{\alpha}\right) + O(k^2). \tag{1.53}$$

The first term on the RHS is the total momentum, whose motion is unaffected by interactions, the second one is the dominant long–wavelength excitation of interest here, which has exactly the same form as the operator $\hat{A}$ defined above. The corresponding form can be understood considering the effect of the affine transformations on the unit sphere, which is illustrated schematically in Figure 1.3. For the longitudinal component of the current, corresponding to $f_{xc}^{L\,\mathrm{dyn}}$, we have e.g. $a_{xx}^L = 1$, other components vanishing; for the transverse component, corresponding to $f_{xc}^{T\,\mathrm{dyn}}$, we have e.g. $a_{xy}^T = 1$, other components vanishing. The usual isotropic compression is described by $a_{ij}^I = \delta_{ij}/d$, and corresponds to the static limit $f_{xc}^{L\,\mathrm{stat}}$.

Notice that the unitary transformation of equation (1.51) tranforms the canonical momentum according to

$$\mathbf{p}_i \to \mathbf{p}_i - \lambda(t)a_{ji}\mathbf{p}_j \tag{1.54}$$

(indeed, $[i\lambda\hat{A}, \mathbf{p}_i^{\alpha}] = -\lambda a_{ji}\mathbf{p}_j^{\alpha}$). Consequently, the Fermi surface of the sheared state is deformed as the unit sphere in real-space: apart from rotations and a phase factor, the plots of Figure 1.3 represent either a real–space sphere embedded in the fluid, or the Fermi surface.

The energy of the deformed Fermi surface, as a function of the deformation parameter $\lambda(t)$ is straightforwarly calculated in terms of Landau parameters

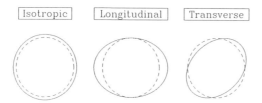

FIGURE 1.3: Shape of the deformed Fermi surface in the three relevant cases. The dashed curves give the original Fermi surface.

(Subsection 1.5.4), whereas the real-space interpretation allows a simple relation with elastic constants (Subsection 1.5.3). These results depend crucially on the assumed unitary character of the time-evolution of the momentum distribution under shear stress, i.e., no collisions must be taking place.

The relation with the low-frequency, dynamic f_{xc} is most simply obtained from the TD-DFT formulation (equations (1.5–1.8)), which gives

$$\Delta E_\lambda^{L,T} - \Delta E_\lambda^{L,T\,\text{ideal}} = \frac{1}{2}\lambda^2 n^2 f_{xc}^{L,T\,\text{dyn}} \tag{1.55}$$

where $\Delta E_\lambda = E_\lambda - E_{\lambda=0}$, and $E_\lambda^{L(T)}$ is the energy of the state with a longitudinal (transverse) deformation of strength λ.

1.5.3   Connection with classical elasticity theory

The stress tensor u_{ij} corresponding to the transformation $\mathbf{r} \to \mathbf{r}'$ is

$$u_{ij} = \frac{1}{2}\left(\frac{\partial \mathbf{r}'_i}{\partial \mathbf{r}_j} + \frac{\partial \mathbf{r}'_j}{\partial \mathbf{r}_i}\right) - \delta_{ij}\,, \tag{1.56}$$

and in an isotropic and uniform system the corresponding energetic cost is [25]

$$\Delta E = \frac{1}{2}\left(K - \frac{2}{d}\mu\right)(u_{ii})^2 + \mu u_{ik}u_{ik} \tag{1.57}$$

where $K = K_T^{-1}$ and μ denote the bulk and shear moduli, respectively. We now give explicit results for the three relevant cases.

- Longitudinal: $u_{xx} = \lambda$, other components vanish. It follows that $\Delta E^L = \frac{1}{2}(K + \xi\mu)\lambda^2$, i.e. $n^2 f_{xc}^{L\,\text{dyn}} = (K_{xc} + \xi\mu_{xc})$, where $\xi = 4/3$ in 3D and $\xi = 1$ in 2D. The suffix $_{xc}$ indicates the xc contribution, defined as usual as the difference between the exact and the ideal gas values.

- Transverse: $u_{xy} = u_{yx} = \lambda/2$, other components vanish. One gets $\Delta E^T = \frac{1}{2}\mu\lambda^2$, i.e. $n^2 f_{xc}^{T\,\text{dyn}} = \mu_{xc}$.

- Isotropic: $u_{xx} = u_{yy} = u_{zz} = \lambda/d$, $\Delta E^I = \frac{1}{2}K\lambda^2$, which corresponds to the "static" limit $n^2 f_{xc}^{L\,\text{stat}} = K_{xc}$.

These results show that $f_{xc}^{L\,\text{dyn}} - \xi f_{xc}^{T\,\text{dyn}} = f_{xc}^{L\,\text{stat}}$, i.e. that the combination $f_{xc}^L - \xi f_{xc}^T$ has the same limit in the two cases examined, since $f_{xc}^{T\,\text{stat}} = 0$ from equation (1.40).

1.5.4   Static Landau theory

The energy functional

$$\Delta E[\delta n_{\mathbf{p}}] = \sum_{\mathbf{p}} \frac{k_F}{m^*}(p - k_F)\delta n_{\mathbf{p}} + \frac{1}{2V}\sum_{\mathbf{pp'}} f_{\mathbf{pp'}}\delta n_{\mathbf{p}}\delta n_{\mathbf{p'}} \tag{1.58}$$

is as usual parametrized in terms of the F_l's, defined in equation (1.48) in 3D and from $f_{\mathbf{pp'}} = (\pi/m^*)\sum_l F_l \cos(l\theta_{\mathbf{pp'}})$ in 2D. Only spin–symmetric terms are relevant here, spin–antisymmetric ones being related in an analogous way to the spin–resolved xc kernels.

The Fermi surface displacement in direction $\hat{\mathbf{p}}$, obtained from equation (1.54), can be parametrized as

$$\delta k_F(\hat{\mathbf{p}}) = k_F|\hat{\mathbf{p}}_j - \lambda a_{ji}\hat{\mathbf{p}}_i| - k_F \simeq -k_F\lambda a_{ji}\hat{\mathbf{p}}_i\hat{\mathbf{p}}_j = k_F v_{\mathbf{p}}. \tag{1.59}$$

in terms of the matrix a_{ij} defined in Section 1.5.2. In the following we use the shorthand notation $v_{\mathbf{p}} = -\lambda a_{ji}\hat{\mathbf{p}}_i\hat{\mathbf{p}}_j$ for the relative displacemente in direction $\hat{\mathbf{p}}$.

Straightforward computations lead to

$$\Delta E[u_{ij}] = n\frac{dk_F^2}{2m^*}\left[\langle v_{\mathbf{p}}^2\rangle + \sum_l F_l\langle P_l(\cos\theta_{\mathbf{pp'}})v_{\mathbf{p}}v_{\mathbf{p'}}\rangle\right] \tag{1.60}$$

where $\langle\ldots\rangle$ denotes angular average over the Fermi surface, $P_l(\cos\theta)$ indicates Legendre polynomials in 3D and equals $\cos(l\theta)$ in 2D.

We now examine the same three cases examined above for the 3D case.

- Longitudinal: $v_{xx} = \lambda$ and other components vanish,

$$\Delta E^L = \frac{\lambda^2}{2}n\frac{k_F^2}{m^*}\left(\frac{3}{5} + \frac{F_0}{3} + \frac{4F_2}{75}\right) \tag{1.61}$$

which gives

$$f_{xc}^{L\,\text{dyn}} = \frac{3}{5}\frac{2E_F}{n}\frac{\frac{5}{9}F_0 - \frac{1}{3}F_1 + \frac{4}{45}F_2}{1 + F_1/3} \tag{1.62}$$

in agreement with the previous result (1.46).

TABLE 1.2: Relative difference between the two limits of the longitudinal f_{xc}, as defined in equation (1.70), as obtained from the estimates of the Landau parameters by Rice [34], Hedin [35], and Yasuhara and Ousaka (YO) [36]. The last column gives the corresponding 2D result obtained from the simulation data on F_l by Kwon, Ceperley and Martin (KCM) [37].

r_s	Rice	Hedin	YO	KCM(2D)
1	–	–	−0.2	−0.03
2	0.07	−0.1	−0.1	0.02
3	–	–	−0.09	0.037
4	0.1	0.006	−0.08	–
5	–	–	−0.08	0.041

- Transverse: $v_{xy} = \lambda$,

$$\Delta E^T = \frac{\lambda^2}{2} n \frac{k_F^2}{m^*} \left(\frac{1}{5} + \frac{F_2}{25} \right) \tag{1.63}$$

which gives

$$f_{xc}^{T\,\mathrm{dyn}} = \frac{2E_F}{5n} \frac{-\frac{1}{3}F_1 + \frac{1}{5}F_2}{1 + F_1/3} \tag{1.64}$$

again, in agreement with the previous result (1.47).

- Isotropic compression: $v_{\mathbf{p}} = \lambda/3$ does not depend on $\mathbf{p}$,

$$\Delta E^I = \frac{\lambda^2}{2} n \frac{k_F^2}{3m^*} (1 + F_0) \tag{1.65}$$

in agreement with the usual Landau theory result for the compressibility, and therefore with the compressibility sum rule for the static limit.

The same computation in 2D gives

$$f_{xc}^{L\,\mathrm{dyn}} = \frac{2\varepsilon_F}{n} \frac{\frac{1}{2}F_0 - \frac{3}{8}F_1 + \frac{1}{8}F_2}{1 + \frac{1}{2}F_1}, \tag{1.66}$$

$$f_{xc}^{T\,\mathrm{dyn}} = \frac{2\varepsilon_F}{n} \frac{-\frac{1}{8}F_1 + \frac{1}{8}F_2}{1 + \frac{1}{2}F_1}, \tag{1.67}$$

$$f_{xc}^{L\,\mathrm{stat}} = \frac{2\varepsilon_F}{n} \frac{\frac{1}{2}F_0 - \frac{1}{4}F_1}{1 + \frac{1}{2}F_1}. \tag{1.68}$$

1.5.5   Discussion and numerical evaluation

A few remarks are in order here:

- If $F_l = 0$ for $l \geq 1$ the two limits of the longitudinal component coincide. This result was first obtained in the 3D case by Lipparini *et al* [38] following a suggestion made by Nozières.

- Without independent knowledge of the Landau parameters, one might argue that $f_{xc}^{L,T}$ are indeed continuous and that the present results constitute a new sum rule on the Landau parameters, stating

$$F_2 = \frac{2d-1}{3}F_1 \, . \tag{1.69}$$

  Numerical evaluations reported below contradict this hypothesis.

- For the presently relevant 3D case the actual numerical value of the parameters F_l is not known in the literature with a satisfactory accuracy. Rough estimates based on various approximate results (see below) indicate a discontinuity around 10%. More precise Quantum Monte Carlo results for the 2D case give the same qualitative picture (see below).

- Since the static value of the transverse component is zero (see equation (1.40)) the combination $f_{xc}^L - \frac{4}{3}f_{xc}^T$ (in 2D, $f_{xc}^L - f_{xc}^T$) is indeed continuous.

- Only Landau parameters up to F_2 enter the result, since in eqn. (1.60) $v_{\mathbf{p}}$ is quadratic in $\hat{\mathbf{p}}$. The fact that $v_{\mathbf{p}}$ is quadratic comes from the fact that the relation between $\mathbf{p}'$ and $\mathbf{p}$ is linear, which is a direct consequence of the $k \to 0$ limit.

A numerical evaluation of the above results can be done using estimates of F_l from the literature. Table 1.2 contains the results for the relative difference of the two limits of the longitudinal component

$$\frac{f_{xc}^{L\,\mathrm{dyn}} - f_{xc}^{L\,\mathrm{stat}}}{f_{xc}^{L\,\mathrm{stat}}} \tag{1.70}$$

obtained in 3D from the values of F_l by Rice [34], Hedin [35] and Yasuhara and Ousaka [36]. Hedin gives no estimate of F_2, we used zero since also in the other two cases the value of F_2 is very small. For small r_s the exact asymptotic behaviours of F_0 and F_1 are known in 3D. Neglecting F_2 the $r_s \to 0$ limit of equation (1.70) is

$$-\frac{8}{5}\left(2 + \ln\frac{\alpha r_s}{\pi}\right) \, . \tag{1.71}$$

In 2D accurate values of the Landau interaction parameters have been obtained from Quantum Monte Carlo simulations by Kwon, Ceperley and Martin

[37]. Their results give reliable estimates for the limits (1.66–1.68), and reject the hypothetical sum rule (1.69), which for $d = 2$ reduces to $F_1 = F_2$. Indeed, from the data in Table VIII of Ref. [37] one gets $F_1 - F_2 = -0.07(3)$, $0.06(3)$, $0.24(3)$, and $0.62(5)$ for $r_s=1$, 2, 3 and 5 respectively, where the digits in parentheses represent the error bar in the last decimal place. The values of the relative discontinuity for the two–dimensional f_{xc}^L given in Table 1.2 are obtained from the same simulation results.

Chapter 2

Two–pair excitations and dynamic exchange–correlation kernels

An approximate decoupling of the equation of motion for the current–current response function allows to obtain a microscopic model for the the dynamical xc kernels $f_{xc}^{L,T}(\omega)$ (Section 2.1, some details are given in Appendix A). The resulting mode–coupling structure indicates a significant role of two–plasmon processes in both f_{xc}^L and f_{xc}^T, which turn out to be very similar. Numerical evaluations (Section 2.2) give evidence for a strong spectral structure near twice the plasma frequency, deriving from the two-plasmon threshold for two-pair excitations [39–41]. This structure is absent in previous estimates, which include perturbation theory and the smooth interpolation scheme of Gross and Kohn (GK) (Section 2.3). Contrary to the GK case, in our scheme the real part is considerably enhanced at intermediate frequencies $\omega \sim \omega_{pl}$ with respect to the static ($\omega = 0$) value.

The present results imply the presence of a third peak in the dynamical structure factor $S(k, \omega)$ at long wavelength, at frequencies higher than both the particle–hole continuum and the plasmonic resonance, which may be observable in inelastic scattering experiments (Figure 2.12).

Our analytical results for f_{xc}^L are finally used to formulate a selfconsistent model for the plasmon dispersion coefficient in alkali metals which emphasizes the role of two plasmon processes (Section 2.4), obtaining good qualitative agreement with available experimental data. Section 2.5 gives a brief summary of the most significant results obtained on $f_{xc}^{L,T}$.

The model can be straightforwardly generalized to 2D [42,43], with similar conclusions which will not be discussed here.

2.1   Approximate decoupling at long wavelength

We present below an approximate decoupling of the equation of motion for the current–current response function and use it to study the long–wavelength exchange–correlation kernels. Our starting point is an exact expression for $\tilde\chi_{ij}(k,\omega)$ to order k^2, which can be obtained by considering only a small number of commutators. This happens because each commutator with the kinetic energy gives an additional power of k, while negative powers of k are absent from the equation of motion after excluding PE^{res}. Indeed, the term $\ll \left[\mathbf{j}_\mathbf{k}^{(i)},H\right];\left[\mathbf{j}_{-\mathbf{k}}^{(j)},H\right]\gg_\omega$ in the equation of motion (1.27) for $\ll \mathbf{j}_\mathbf{k}^{(i)};\mathbf{j}_{-\mathbf{k}}^{(j)}\gg$ can be written (see Appendix A) in terms of four–point response functions $\ll AB;CD\gg_\omega$, where A, B, C and D are either $\rho_\mathbf{k}$ or $\mathbf{j}_\mathbf{k}$ operators. In turn, the four–point response functions can be approximately decoupled with the following RPA–like ansatz:

$$\langle A(t)B(t)C(0)D(0)\rangle \simeq \langle A(t)C(0)\rangle\langle B(t)D(0)\rangle + \langle A(t)D(0)\rangle\langle B(t)C(0)\rangle \quad (2.1)$$

which in frequency domain corresponds to

$$\mathbf{Im}\,\ll AB;CD\gg_\omega \;\simeq\; -\int_0^\omega \frac{d\omega'}{\pi}\,[\mathbf{Im}\,\ll A;C\gg_{\omega'}\mathbf{Im}\,\ll B;D\gg_{\omega-\omega'}$$
$$+\mathbf{Im}\,\ll A;D\gg_{\omega'}\mathbf{Im}\,\ll B;C\gg_{\omega-\omega'}]\,. \quad (2.2)$$

Such a decoupling includes by construction two–pair processes, which are the lowest order processes with nonzero spectral strength in the relevant region of the (k,ω) plane. Perturbative computations evaluate the LHS of equation (2.2) for an ideal gas and obtain a spectrum restricted to single particle excitations, which is essentially equivalent to the one obtained using noninteracting response functions in the RHS (the difference lies in exchange processes, as discussed below). We intend to include the effect of plasmons, and therefore shall use RPA response functions in the RHS.

Quite lengthy computations, described in some detail in Appendix A, lead to the following result for the long–wavelength exchange–correlation kernels $f_{xc}^{L,T}(\omega)$:

$$\mathbf{Im}\,f_{xc}^{L,T}(\omega) \;=\; -\int_0^\omega \frac{d\omega'}{\pi}\int \frac{d^3q}{(2\pi)^3 n^2} v_q^2 \frac{q^2}{(\omega-\omega')^2}\mathbf{Im}\,\chi_L(q,\omega-\omega')$$
$$\times\left[a_{L,T}\frac{q^2}{\omega'^2}\mathbf{Im}\,\chi_L(q,\omega') + b_{L,T}\frac{q^2}{\omega^2}\mathbf{Im}\,\chi_T(q,\omega')\right] \quad (2.3)$$

with $a_L = 23/30$, $a_T = 8/15$, $b_L = 8/15$ and $b_T = 2/5$. The expression for the longitudinal part is equivalent to the one obtained by Hasegawa and Watabe [44] by diagrammatic means, and similar – though not identical, see Section A.2.4 – to the one obtained by Neilson *et al* [45]. Related expressions were obtained within a perturbative formalism by several authors [46–53]. The result for the transverse component is new. Analogous results in 2D have been recently obtained by R. Nifosì [42], and differ from the above only in the coefficients $a_{L,T}$ and $b_{L,T}$

Equation (2.3) describes composite excitations, consisting in two elementary excitations (either single pairs or plasmons). In particular, the χ_L-χ_L term contains double plasmon excitations, which are hereby predicted to play a role in the linear response, both longitudinal and transverse. Such contributions give nonvanishing contributions to **Im** χ at long wavelength and arbitarily high frequency (see below).

As mentioned above, equation (2.3) neglects exchange processes. Considering the excitation of pairs out of the Fermi sphere it can be seen that exchange is expected to reduce the total result by a factor of 2 when ω is large on the scale of the Fermi energy*. We incorporate an approximate treatment of exchange in our results multiplying **Im** f_{xc} by

$$g_x(\omega) = \frac{\beta + 0.5\omega/2\varepsilon_F}{1 + \omega/2\varepsilon_F} \qquad (2.4)$$

where β is an additional parameter determined as explained below, which would be 1 in a purely perturbative approach. Including this correction we are able to essentially reproduce the perturbative results replacing the response functions on the RHS of equation (2.3) with noninteracting response functions.

The real part of f_{xc} can be immediately obtained by Kramers-Kronig

$$\mathbf{Re}f_{xc}(\omega) = \mathbf{Re}f_{xc}(\infty) + \frac{1}{\pi}\int_{-\infty}^{+\infty} d\omega' \frac{\mathbf{Im}f_{xc}(\omega')}{\omega' - \omega}. \qquad (2.5)$$

and the high–frequency limits given in equations (1.37–1.38).

For $\omega = 0$ equation (2.5) gives an estimate for the low–frequency limit, which has been shown to be discontinuous in Section 1.5. If taken literally, this model would also indicate a discontinuity of approximately the same magnitude. Due to (i) the limited estimated discontinuity, (ii) the uncertainty in the precise value of the discontinuity itself, and (iii) the appeal of a theory which reduces continuously to LDA as $\omega \to 0$, we prefer instead to enforce continuity in our f_{xc}^L. This is done by fitting the parameter β entering equation (2.4) in order to reproduce the compressibility sum rule result (equation (1.39)) for $\omega = 0$. Note that the final values of β turn out to be close to 1 at metallic densities (see Table 2.1). For consistency the same factor $g_x(\omega)$ is used in the transverse component.

The high frequency limit of the above expressions can be obtained analytically, since for large k and ω the Single Pole Approximation (SPA, see Sec.

---

*Such a statement can be made rigorous within second–order perturbation theory.

2.2.1) is asymptotically exact. In particular,

$$\mathbf{Im}\,\chi_{L,T}(k,\omega) \simeq -\frac{\pi}{2}\frac{nh_{L,T}}{m}\left[\delta\left(\omega - \frac{k^2}{2m}\right) - \delta\left(\omega + \frac{k^2}{2m}\right)\right] \qquad (2.6)$$

where the sum rule (1.30) gives $h_L = k^2/2m$ and $h_T = 4\langle ke\rangle/3$. The fact that $\mathbf{Im}\,\chi_T$ is smaller than $\mathbf{Im}\,\chi_L$ by a factor of k^2 at large k and ω is also evident from equation (B.4) and implies that only the first (L–L) term of equation (2.3) contributes for high ω. Direct substitution and integration gives

$$\mathbf{Im}\,f_{xc}^{L,T}(\omega \to \infty) \simeq -2c_{L,T}\left(\frac{2Ry}{\omega}\right)^{3/2}a_0^3 Ry \qquad (2.7)$$

where $c_L = 23\pi/15$ and $c_T = 16\pi/15^\dagger$, in agreement with the result of Glick and Long [48] for the longitudinal term and its extension to the transverse one [42]. Kramers–Kronig then proves that to leading order in $\omega \to \infty$ the same behaviour is found in the real part, $\mathbf{Re}\,f_{xc}(\omega) \simeq f_{xc}(\infty) - \mathbf{Im}\,f_{xc}(\omega)$.

2.2 Numerical results for f_{xc}

2.2.1 Single Pole Approximation

In this Section we obtain the first quantitative results for the xc kernels f_{xc} neglecting the $\chi_L - \chi_T$ term in the RHS side of equation (2.3) and using the Single Pole Approximation (SPA) for the longitudinal response function. The main qualitative inaccuracy of the SPA is the absence of the low–frequency particle–hole excitations, and is overcome by the RPA computation described in the following Section. The SPA is indeed more appropriate for the bosonic plasma, where it is the outcome of a vast class of Feynman–like theories [54].

In the SPA the entire excitation spectrum is collapsed in a single collective excitation, which we shall call "plasmon" at any k, although strictly speaking such a name is appropriate only at long wavelength. The dispersion relation ω_k and the f–sum rule determine the longitudinal response function

$$\mathbf{Im}\,\chi_L(k,\omega) \simeq -\frac{\pi}{2}\frac{n\omega_k}{m}\left[\delta\left(\omega - \omega_k\right) - \delta\left(\omega + \omega_k\right)\right]. \qquad (2.8)$$

A further simplification is obtained by adopting a parametrized form for the plasmon dispersion,

$$\omega_k = \sqrt{\omega_{pl}^2 + 4\alpha\varepsilon_k\omega_{pl} + \varepsilon_k^2} \qquad (2.9)$$

where $\varepsilon_k = k^2/2m$ is the noninteracting kinetic energy, which is expected to agree with the interacting one at wavelengths much smaller than the average

---

†in atomic units $m = e = \hbar = 1$, $\mathbf{Im}\,f_{L,T}^{xc} \simeq -c_{L,T}\omega^{-3/2}$

interparticle spacing. Equation (2.9) interpolates between small and large k behaviours and allows to keep account of the variation of the dispersion curve with density through the dispersion coefficient α. At small r_s α is positive and ω_k monotonically increasing, whereas at large r_s (larger than about 6) α and negative at ω_k exhibits a minimum at intermediate k.

We have shown above that $\mathbf{Im}\, f_{xc}^{L,T}(\omega)$ vanishes as $-\omega^{-3/2}$ for high ω. As a consequence of the SPA and of the mode–coupling structure of the present approximation it equals zero at frequencies below $2\omega_{min}$, ω_{min} being the minimum frequency attained by the plasmon dispersion curve. In the intermediate range we expect a minimum which reflects directly the role of two–plasmon processes.

For positive α a continuous function with an infinite right–side derivative at $2\omega_{min}$ is obtained (left panel of Figure 2.1), whereas for negative α a divergence of the type $(\omega - 2\omega_{min})^{-1/2}$ is found (right panel of Figure 2.1). This in turn leads to a cusp in $\mathbf{Re}\, f_{xc}^{L}(\omega)$ for positive α and to a singularity of the type $(2\omega_{min} - \omega)^{-1/2}\,\theta(2\omega_{min} - \omega)$ if α is negative (Figure 2.1). This behaviour can be tuned artificially, by using different values of α at each r_s (see Figure 2.2). In all cases $\mathbf{Re}\, f_{xc}^{L}(k,\omega)$ at low frequency lies below the value $f_{xc}^{L}(0)$ given by the compressibility sum rule and the limiting value $f_{xc}^{L}(\infty)$ corresponding to the third moment is reached from above. While the above–mentioned singularities at $2\omega_{min}$ are consequences of the SPA, singularities in the two–pair spectrum at $2\omega_{pl}$ arise even in the RPA (see below). In general we remark that the sharp feature at $\omega \simeq 2\omega_{min}$ arises from phase–space effects, and therefore we do not expect it to be completely washed out in a more refined theory, even if quantitative values can be significantly affected.

We do not discuss $f_{xc}^{T}(\omega)$ separately since we are neglecting transverse contributions to equation (2.3) and therefore

$$\mathbf{Im}\, f_{xc}^{T}(\omega) = \frac{16}{23}\mathbf{Im}\, f_{xc}^{L}(\omega)\,. \qquad (2.10)$$

In the case of bosons an upper bound on the plasmon dispersion coefficient has been demonstrated from an exact sum rule argument [55], giving $\alpha \leq 1/2\omega_{pl}nK_T$, where K_T is the compressibility (see equation (1.39), in the case of bosons $1/K_T^0 = 0$). In terms of f_{xc}^{L}, this upper bound states $\mathbf{Re}\, f_{xc}^{L}(\omega_{pl}) \leq \lim_{k\to 0} f_{xc}^{L}(k,\omega = 0)$, and therefore indicates that $\mathbf{Re}\, f_{xc}^{L}(\omega_{pl})$ lies below $f_{xc}^{L}(0)$. In turn, this implies the presence of a minimum at finite ω, since $f_{xc}^{L}(\infty)$ lies above $f_{xc}^{L}(0)$. A structure in the frequency dependence of $\mathbf{Re}\, f_{xc}^{L}(\omega)$ of the type that we have found is thus correct for the boson plasma at all couplings. No such exact statement can be made at present for fermions. In the following we deal with evaluations of $f_{xc}^{L,T}(\omega)$ from equation (2.3) performed with more refined approximations for the response functions appearing in the LHS, which will confirm this basic result.

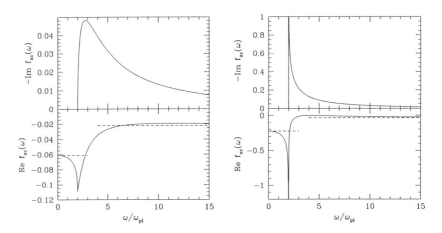

FIGURE 2.1: Real and imaginary parts of $f_{xc}^{L}(k,\omega)$ at small k in units of $2\omega_{pl}/n$, as functions of ω/ω_{pl} from the SPA at $r_s = 1$ (left panel) and $r_s = 10$ (right panel). The asymptotic behaviours are shown by the dashed lines. The plasmon dispersion coefficients α are as determined in Section 2.4.

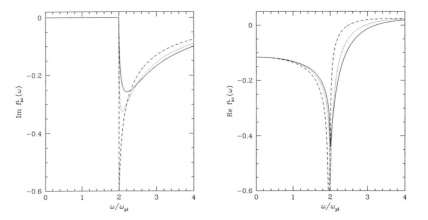

FIGURE 2.2: Imaginary (left panel) and real (right panel) part of $f_{xc}^{L}(\omega)$ as functions of ω/ω_{pl} at $r_s = 5$ computed in the SPA with $\alpha = 0.02$ (full curves), 0.01 (dotted curves) and -0.01 (dashed curves).

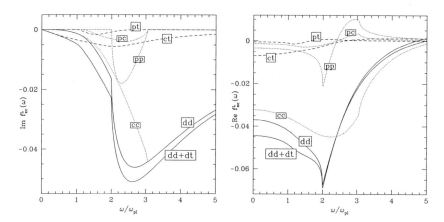

FIGURE 2.3: Imaginary (left panel) and real (right panel) parts of $f_{xc}^L(\omega)$, in units of $2\omega_{pl}/n$, as functions of ω/ω_{pl} at $r_s = 1$. The dotted curves give the three dd components (cc, pc and pp), the dashed curves the two dt components (ct and pt), the full curves the total (dd+dt) and the partial sum of the three dd terms. The constant $f_{xc}^L(\infty)$ and the scaling function $g_x(\omega)$ have been omitted for simplicity.

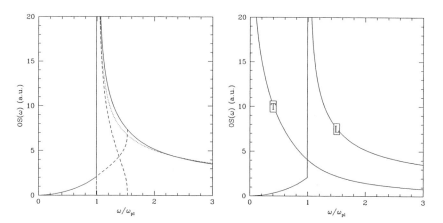

FIGURE 2.4: Left panel: Momentum–integrated oscillator strength at $r_s = 1$ as a function of ω/ω_{pl} in the RPA (full curve) and SPA (dotted curve). The two dashed curves give the plasmon and continuum contributions to the RPA result. Right panel: comparison between the total oscillator strength of the longitudinal (L) and transverse (T) components of the RPA response function.

2.2.2   Random Phase Approximation

This Section presents the numerical results obtained using the RPA to evaluate
the response functions in the LHS of equation (2.3). When retardation is
neglected ($c \to \infty$) the RPA transverse response function reduces to the ideal
Fermi gas one [27], which is computed explicitly in Appendix B. For simplicity
in the discussion of the separate components we neglect the factor $g_x(\omega)$ and
concentrate on f_{xc}^L. The real parts are obtained by application of Kramers–
Kronig to the individual components of the imaginary part, neglecting the
constant $f_{xc}(\infty)$.

We have already seen from equation (2.3) that $f_{xc}(\omega)$ involves a density–
density (dd) contribution and a density–transverse current (dt) contribution
("density" is here equivalent to "longitudinal current"). In turn, in the RPA
the longitudinal spectrum contains a strong plasmon resonance and a broad
particle–hole continuum, whereas the transverse one only contains the continuum. We are now going to analyze the different contributions deriving from
those terms.

The dd term involves plasmon–plasmon (pp), plasmon–continuum (pc) and
continuum–continuum (cc) components. The pp contribution has a behaviour
around frequency $2\omega_{pl}$ very similar to the SPA result discussed above. As can
be seen from Figure 2.3, the individual components show additional singularities at ω_c and $2\omega_c$, ω_c being the frequency where the plasmon dispersion
curve crosses the edge of the single–pair continuum. However, these singularities cancel away in the sum. On the other hand, the singularities at ω_{pl} and
$2\omega_{pl}$ arise from real phase–space effects and are present in the components as
well as in their sum. Section 2.2.3 gives an analytic treatment of the singularity which is brought by the pp spectral component into the real part of the
exchange–correlation kernel at $2\omega_{pl}$.

The dt contribution involves two components, i.e. (i) a ct component arising from a single–pair density excitation and a single–pair transverse excitation
and (ii) a pt component arising from a plasmon and a single–pair transverse
excitation. We remark that the dt term is dominant over the dd one in the
low frequency region. The qualitative comparison between the various contributions can be understood in terms of the wavevector–integrated oscillator
strengths

$$OS^{L,T}(\omega) = \int d^3\mathbf{k}\, \omega\, v_k\, \frac{k^2}{\omega^2} \mathbf{Im}\, \chi^{L,T}(k,\omega) \tag{2.11}$$

plotted in Figure 2.4. The longitudinal oscillator strength is concentrated
around frequency $2\omega_{pl}$ due to the strong plasmon resonance and vanishes as ω^2
at low frequency, leading to a ω^3 behaviour in the dd contribution to f_{xc}. The
transverse OS^T instead has a logarithmic divergence at low frequency, which
leads to a linear behaviour in the dt contribution to $\mathbf{Im}\, f_{xc}^L$ (see Figure 2.3).
The oscillator strength residing in the transverse excitations at low frequency

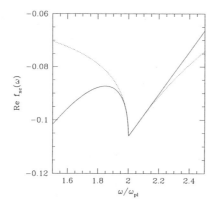

FIGURE 2.5: $\mathbf{Re}\, f_{xc}(\omega)$ as obtained numerically around $2\omega_{pl}$ at $r_s = 1$ in the SPA (dotted curve), compared with the singular behaviour given in equation (2.16) added to a straight line (full curve).

induces through the pt component some structure in $\mathbf{Re}\, f_{xc}(\omega)$ in the region of the plasma frequency, as can be seen from Figure 2.3.

2.2.3   Singularity in $\mathbf{Re}\, f_{xc}(\omega)$ at $2\omega_{pl}$

In this Section the non–smooth (non–C^1) terms in $\mathbf{Re}\, f_{xc}(\omega)$ around $2\omega_{pl}$ are evaluated analytically in the case $\alpha > 0$. It is also verified that the result agrees with the numerical evaluation reported above. In the case $\alpha < 0$ a behaviour of the type $(\omega - \omega_{pl})^{-1/2}$ is easily derived as pointed out in Section 2.2.1.

The non–smooth contribution to $\mathbf{Im}\, f_{xc}(\omega)$ due to the two–plasmon channel is

$$\mathbf{Im}\, f_{xc}(\omega \sim 2\omega_{pl}) \propto \theta(\omega - 2\omega_{pl})\sqrt{\omega - 2\omega_{pl}} \qquad (2.12)$$

apart from an overall negative constant factor. When the real part is obtained by the Kramers–Kronig relation (2.5) the only singular contributions arise for $\omega' = 2\omega_{pl}$, so that neglecting smooth terms the integration range can be limited to $[2\omega_{pl}, 2\omega_{pl} + \Delta]$. By the changes of variable $\omega' = 2\omega_{pl} + x$ and $\omega = 2\omega_{pl} + y$ the integral can be put into the form

$$\int_{2\omega_{pl}}^{2\omega_{pl}+\Delta} d\omega' \frac{\sqrt{\omega' - 2\omega_{pl}}}{\omega' - \omega} = \int_0^\Delta dx \frac{x^{1/2}}{x - y} \quad . \qquad (2.13)$$

If $y > 0$, i.e. for $\omega > 2\omega_{pl}$ we have

$$\int_0^\Delta \frac{x^{1/2}}{x - y}dx = 2\Delta^{1/2} - y^{1/2} \ln\left|\frac{y^{1/2} + \Delta^{1/2}}{y^{1/2} - \Delta^{1/2}}\right| \simeq 2\Delta^{1/2} - \frac{2}{\Delta^{1/2}}y \qquad (2.14)$$

TABLE 2.1: Exact limiting behaviours of $f_{xc}^{L,T}(\omega)$ from the Monte Carlo equation of state, in units of $2\omega_{pl}/n$, and best fit parameters for $\mathbf{Im}\ f_{xc}^{L}$ according to (2.17).

r_s	$f_{xc}^{L}(0)$	$f_{xc}^{L}(\infty)$	$f_{xc}^{T}(\infty)$	β	$10^2 c_0$	$10^2 c_1$	ω_1	ω_2	d_0	$10^2 d_1$
0.5	-0.0425	-0.0183	0.0175	1.84	0.179	0.71	1.77	-3.71	0.173	6.01
1	-0.0611	-0.0222	0.0280	1.44	0.421	1.76	0.982	-1.45	0.291	9.38
2	-0.0889	-0.0253	0.0453	1.21	0.895	3.87	0.347	0.181	0.49	13.2
3	-0.1113	-0.0265	0.0603	1.12	1.3	5.9	0.034	0.921	0.664	15.3
4	-0.1309	-0.0268	0.0739	1.08	1.64	7.95	-0.139	1.32	0.824	17.1
5	-0.1487	-0.0271	0.0864	1.06	1.94	9.83	-0.267	1.61	0.974	18.3
6	-0.1653	-0.0274	0.0980	1.04	2.19	11.7	-0.364	1.82	1.12	19.3
10	-0.2240	-0.0331	0.1366	0.926	3.06	18	-0.577	2.29	1.64	22
15	-0.2883	-0.0535	0.1705	0.679	3.83	24.5	-0.702	2.56	2.22	23.8
20	-0.3477	-0.0869	0.1923	0.336	4.69	28.9	-0.771	2.72	2.75	24.5

while if $y < 0$, i.e. for $\omega < 2\omega_{pl}$ we have

$$\int_0^\Delta \frac{x^{1/2}}{x-y}dx = 2\Delta^{1/2} - 2|y|^{1/2}\tan^{-1}\left(\frac{\Delta^{1/2}}{y^{1/2}}\right) \simeq 2\Delta^{1/2} - \pi|y|^{1/2} - \frac{2}{\Delta^{1/2}}y \quad . \tag{2.15}$$

We thus conclude that a singular term is present only on the left of the singularity and is proportional to $-\theta(2\omega_{pl} - \omega)\sqrt{2\omega_{pl} - \omega}$.

Recovering the multiplicative constant this gives

$$\mathbf{Re}\ f_{xc}(2\omega_{pl} - \delta) = -\mathbf{Im}\ f_{xc}(2\omega_{pl} + \delta) + \text{smooth function (class } C^1) \tag{2.16}$$

which is compared in Figure 2.5 with the result of the numerical evaluation of $\mathbf{Re}\ f_{xc}(\omega)$ within the SPA, the smooth function being represented by a straight line fitted to the numerical result at $\omega > 2\omega_{pl}$. The same applies to the RPA computation.

2.2.4   Analytic interpolation

We found that our numerical results for $\mathbf{Im}\ f_{xc}^{L}(\omega)$ presented in Section 2.2.2 can be accurately reproduced by the expression

$$\mathbf{Im}\ f_{xc}^{L}(\omega) = -g_x(\omega)\begin{cases} c_0\omega + c_1\dfrac{\omega-1}{e^{(7/\omega)-5}+1} & (\omega < 2) \\[2ex] \dfrac{d_0\sqrt{\omega-2}+d_1}{\omega(\omega-\omega_1\sqrt{\omega}-\omega_2)} & (\omega > 2) \end{cases} \tag{2.17}$$

where ω is in units of ω_{pl}, f_{xc}^{L} in units of $2\omega_{pl}/n$, and $g_x(\omega)$ was defined in (2.4). The optimal fit parameters given in Table 2.1 were obtained imposing

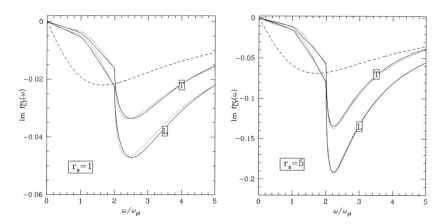

FIGURE 2.6: Imaginary parts of $f_{xc}^L(\omega)$ and $f_{xc}^T(\omega)$ in units of $2\omega_{pl}/n$, as functions of ω/ω_{pl} at $r_s = 1$ (left panel) and $r_s = 5$ (right panel). The present results (full curves) are compared with the Gross–Kohn interpolation for the longitudinal component (dashed curves) and with the fit discussed in Section 2.2.4 (dotted curves).

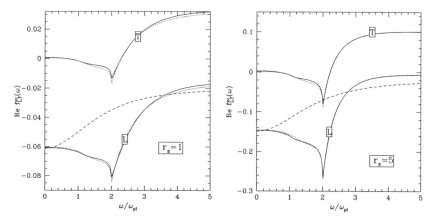

FIGURE 2.7: Real parts of $f_{xc}^L(\omega)$ and $f_{xc}^T(\omega)$ (full curves), in units of $2\omega_{pl}/n$, as functions of ω/ω_{pl} at $r_s = 1$ (left panel) and $r_s = 5$ (right panel). The meaning of the other curves is as in Figure 2.6.

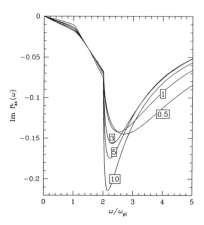

FIGURE 2.8: Imaginary part of $f_{xc}^L(\omega)$ as a functions of ω/ω_{pl} at $r_s = 0.5, 1, 3, 5$ and 10 rescaled in order to fit on the same plot.

(i) continuity at $\omega = 2\omega_{pl}$, (ii) the conservation of the normalization integral $\int \mathbf{Im}\, f_{xc}\, d\omega/\omega$, (iii) the asymptotic behaviour given by equation (2.7), and fitting the remaining 3 parameters to the numerical data.

The real part can be obtained using the Kramers–Kronig relation (equation (2.5)) and the high–frequency value given in Table 1.1; the low–frequency value from the same Table can be used as a check. The transverse spectrum is rather accurately reproduced at all frequencies by setting

$$\mathbf{Im}\, f_{xc}^T(\omega) \simeq 0.72 \cdot \mathbf{Im}\, f_{xc}^L(\omega)\,, \qquad (2.18)$$

whereas for the real part there is an additional shift due to the different $\omega = \infty$ value. The good quality of the resulting fit is shown in Figures 2.6 and 2.7.

2.3   Discussion

Gross and Kohn (GK) proposed [11–13] the following smooth interpolation for $\mathbf{Im}\, f_{xc}^L$:

$$\mathbf{Im}\, f_{xc}^{L\,(GK)}(\omega) = -\frac{a\omega}{(1 + b\omega^2)^{5/4}} \qquad (2.19)$$

where a and b are fixed by imposing the asymptotic behaviour (2.7) and the compressibility limit (1.39) for the real part, obtained from (2.19) through Kramers–Kronig (2.5). The first presentation of (2.19) by Gross and Kohn [11,12] neglected the term $\langle ke \rangle - \langle ke \rangle^0$ in the high–frequency limit (1.37) entering equation (2.5), this inaccuracy was corrected in a later paper by Iwamoto and Gross [13]. Only the corrected version is considered here.

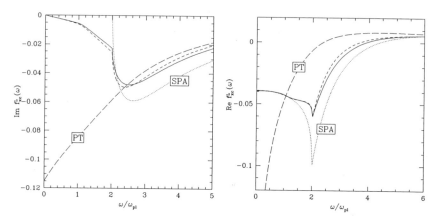

FIGURE 2.9: Imaginary (left panel) and real (right panel) parts of $f_{xc}^L(\omega)$ in units of $2\omega_{pl}/n$, as functions of ω/ω_{pl} at $r_s = 1$. The RPA results (full curves) are compared with the SPA (dotted curves), the STLS (dashed curves) and the perturbative ones (long–dashed curves, PT).

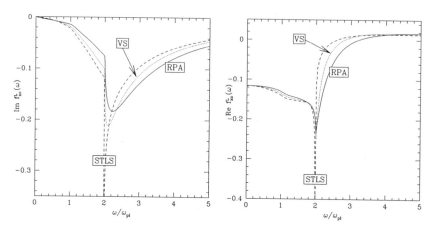

FIGURE 2.10: Imaginary (left panel) and real (right panel) parts of $f_{xc}^L(\omega)$ in units of $2\omega_{pl}/n$, as functions of ω/ω_{pl} at $r_s = 5$. The RPA results (full curves) are compared with the STLS ones (dashed curves) and the VS ones (dotted curves).

Figures 2.6 and 2.7 report our final results for the imaginary and real parts of f_{xc} and compare them with the GK interpolation scheme for the longitudinal term. Both curves reproduce the asymptotic limits (1.37) and (1.39) as well as the $\omega^{-3/2}$ high–frequency behaviour (2.7), but the behaviours at intermediate frequencies are strikingly different. As discussed above our curves for $\mathbf{Re}\, f_{xc}$ exhibit a sharp minimum around $2\omega_{pl}$, which corresponds to the sharp structure found in $\mathbf{Im}\, f_{xc}$ at the same frequency. The physical origin lies in the large spectral strength of the plasmon excitation as compared to single–pair excitations, which accumulates most of the spectral strength of two–pair processes near $2\omega_{pl}$. This spectral structure becomes sharper with increasing coupling strength, i.e. with increasing r_s (see Figure 2.8).

The transverse spectrum turns out to be very similar to the longitudinal one. From equation (2.3) and the discussion of the various components outlined in Section 2.2.2 it can be seen that $\mathbf{Im}\, f^T_{xc} = 16/23 \cdot \mathbf{Im}\, f^L_{xc}$ for $\omega \gg \omega_{pl}$ whereas $\mathbf{Im}\, f^T_{xc} = 3/4 \cdot \mathbf{Im}\, f^L_{xc}$ for $\omega \ll \omega_{pl}$. This gives a formal justification to equation (2.18). We further note that within the accuracy of the present model $f^T_{xc}(\omega = 0)$ is indistinguishable from zero in the entire density range explored here.

The second order perturbative results, as noticed above, can be essentially reproduced by using ideal gas response functions in the LHS of (2.3). Figure 2.9 shows that the high frequency behaviour is almost unchanged, but the low frequency behaviour is completely different due to the absence of the plasmon and the overestimate of the role of single pairs in perturbation theory.

A further assessment of the reliability of the present results can be obtained by comparing computations done with a more refined response function in the LHS of equation (2.3). Figure 2.9 compares the present results at $r_s = 1$ with analogous computations performed by R. Nifosì using the response function given by the Singwi–Tosi–Land–Sjölander (STLS) [56] scheme, and shows that at this low coupling strength there is no appreciable difference. The analogous comparison at $r_s = 5$ is displayed in Figure 2.10. In this case the dispersion coefficient in the STLS approximation is (slightly) negative, giving rise to a divergence in $\mathbf{Im}\, f_{xc}$ at $\omega = 2\omega_{pl}$. The Vashishta–Singwi (VS) [57] response function has a dispersion coefficient still positive, but lower than the RPA, and therefore has an intermediate behaviour. The main consequence of the introduction of the static local field correction is that the plasmon dispersion coefficient in the response functions in the RHS of equation (2.3) is negative, and this in turn mainly affects the the plasmon-plasmon (pp) part of the density-density (dd) term, in the language of Section 2.2.2. The qualitative effect is the same as obtained artificially changing the dispersion coefficient in the SPA computation (see Figure 2.2 and discussion in Section 2.2.1).

We finally turn to examine some consequences of the present results for $f^{L,T}_{xc}(\omega)$, also in connection with experimentally measurable quantities. By Fourier transforming the data for $f_{xc}(\omega)$ we can obtain the time–dependent xc

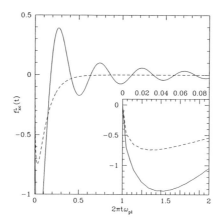

FIGURE 2.11: Longitudinal f_{xc}^L as a function of time (in units of $2\pi/\omega_{pl}$) at $r_s = 5$. The present results (full curve) are compared with the GK interpolation (dashed curve). The inset shows an enlargement of the small time part.

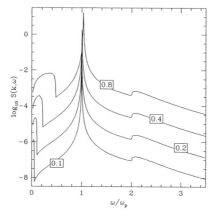

FIGURE 2.12: Dynamic structure factor $S(k,\omega)$ at $r_s = 5$ as a function of ω/ω_{pl} at various values of $k r_s a_B$ on a semilogarithmic scale.

kernel $f_{xc}(t)$:

$$f_{xc}(t) = f_{xc}(\omega = \infty)\delta(t) + f_{xc}^{\text{reg}}(t) \tag{2.20}$$

where the regular component $f_{xc}^{\text{reg}}(t)$ vanishes for $t \leq 0$. Note that in this language the ALDA (equation (1.2)) amounts to $f_{xc}(t) = f_{xc}(\omega = 0)\delta(t)$ and that $\int f_{xc}^{\text{reg}}(t)dt = f_{xc}(\omega = 0) - f_{xc}(\omega = \infty)$. The longitudinal component of $f_{xc}^{\text{reg}}(t)$, reported in Figure 2.11, exhibits strong oscillations at frequency $2\omega_{pl}$ decaying as $t^{-3/2}\cos(\omega_{pl}t/\pi)$ for large t, which are completely absent in the GK curve. The present result for $f_{xc}(t)$ has a considerably longer extension in time, implying xc effects are more persistent in time than in the GK approximation.

Figure 2.12 shows on a semilogarithmic scale the dynamic structure factor

$$S(k, \omega) = -\frac{2k^2}{n\omega^2}\text{Im}\,\chi_L(k, \omega) \tag{2.21}$$

which is relevant to inelastic scattering experiments, computed at small wavevector neglecting the dependence of $f_{xc}^L(k, \omega)$ on k. The threshold behaviour at frequency $2\omega_{pl}$ is a clearcut signature of the role of two plasmon processes in dynamic xc effects, and the linewidth of the long–wavelength plasmon reflects decay into two particle–hole pairs. Further consequences on plasmon dispersion are discussed in some detail in the next Section.

Recent experiments of inelastic X-ray scattering on Aluminum allowed to extract interesting information on the longitudinal component $f_{xc}^L(k, \omega)$. In Ref. [23] the authors assume that $f_{xc}^L(k, \omega)$ has a negligible frequency dependence in the experimentally relevant region and use its value at a given wavevector as a fitting parameter to the experimental data. A standard static LDA computation is used to take into account band–structure effects. Their result is significantly higher than the accurate Monte Carlo evaluations of $f_{xc}^L(k, \omega = 0)$ of Moroni, Senatore and Ceperley [58], implying an enhancement of f_{xc}^L in the dynamic regime with respect to the static limit. While quantitative comparison is difficult due to the finite wavevector used in the experiments and to the number of approximations involved, the qualitative trend is in agreement with our prediction.

2.4   Selfconsistent calculation of the plasmon dispersion coefficient

2.4.1   Introduction

From electron energy loss experiments at high resolution vom Felde, Sprösser-Prou and Fink [21] have reported new and accurate data on the plasmon excitation in the alkali metals from Na to Cs. The dispersion relation

$$\omega_k = \omega_{k=0} + \alpha\frac{k^2}{m} + Ck^4 \tag{2.22}$$

can be fitted to their data up to the critical wave number for the onset of Landau damping. Most interestingly the value of the leading dispersion coefficient α drops rapidly through the series of metals, becoming essentially zero in Rb and negative in Cs. While such behaviour of α with decreasing electron density had been qualitatively predicted already in theories of the homogeneous electron fluid developed since the late sixties [56], the observed decrease is much more pronounced than in the corresponding theoretical results [30,57,59–61].

A number of attempts have been made to account for these observations [38,62–67]. By an *ad hoc* matching of the coupling strength for degenerate electrons onto that of the one–component classical plasma Kalman *et al* [65] claimed that dynamic correlation effects in the homogeneous fluid can account for the observed softening of the plasmon dispersion. Further evidence for this view has come from the work of Lipparini *et al* [38] evaluating the plasmon contribution to the compressibility and f-sum rules. In contrast Aryasetiawan and Karlsson [67] have emphasized the role of band structure effects within a random phase approximation (RPA) calculation, with special regard to the influence of low–lying d states coming close to the Fermi surface in the heavier alkali metals. It can in fact be expected that both correlation and band structure effects are relevant for a quantitative account of the observations. A number of attempts based on local and nonlocal time–dependent density functional approaches [66,68,69] within existing theories of exchange and correlation led to a range of theoretical values of α which extends down to the measured values.

Further clarification of the relevant dynamic mechanisms operating in the homogeneous electron gas seems useful. We now present a study which elucidates the role of two–pair excitations by calculating selfconsistently the plasmon dispersion coefficient, the coupled modes being estimated within a single–pole approximation (SPA) and selfconsistency being imposed via the compressibility and/or the third spectral moment sum rule. Our work is based on the first part of equation (2.3), which only involves longitudinal response functions and can be rewritten in terms of the density–density response function alone.

2.4.2   Selfconsistent model

We compute the xc kernel using the approximation presented in Section 2.1, within the simplified SPA scheme discussed in Section 2.2.1. In this case, the expression (2.3) for $\mathbf{Im}\, f_{xc}^{L}$ reduces to

$$\mathbf{Im}\, f_{xc}^{L}(k,\omega) = -g_x(\omega)\frac{23}{30}\frac{\pi\omega_{pl}^2}{(2\pi)^3 n^2}\int d^3q\,\delta\left(\omega - 2\omega_q\right)\frac{\omega_{pl}^2}{\omega^2}. \tag{2.23}$$

The function $g_x(\omega)$ was defined in equation (2.4) and accounts for exchange in the fermionic plasma. For bosons, $g_x(\omega) = \beta$. The real part is then computed via Kramers–Kronig, either using the static limit from equation (1.39) and

$$\mathbf{Re} f_{xc}^L(\omega) = \mathbf{Re} f_{xc}^L(0) + \frac{1}{\pi} \int_{-\infty}^{+\infty} d\omega' \mathbf{Im} f_{xc}^L(\omega') \left(\frac{1}{\omega' - \omega} - \frac{1}{\omega'} \right) \qquad (2.24)$$

or using the high–frequency limit from equation (1.37) and

$$\mathbf{Re} f_{xc}^L(\omega) = \mathbf{Re} f_{xc}^L(\infty) + \frac{1}{\pi} \int_{-\infty}^{+\infty} d\omega' \frac{\mathbf{Im} f_{xc}^L(\omega')}{\omega' - \omega} . \qquad (2.25)$$

As noticed above, the definition of β brings in agreement equation (2.24) and (2.25), for completeness we shall also present results obtained using $\beta = 1$ and choosing either one of the two Kramers–Kronig relations. The plasmon dispersion coefficient α, defined in (2.22), is then determined via

$$\mathbf{Re} f_{xc}^L(k \to 0, \omega_{pl}) = \frac{2\omega_{pl}}{n}(\alpha - \alpha_{\mathrm{RPA}}) . \qquad (2.26)$$

where the RPA value α_{RPA} is zero for bosons and equals $3\varepsilon_F/5\omega_{pl}$ for fermions, ε_F being the Fermi energy.

Since we are mainly interested in the plasmon dispersion at small k, we adopt for the mode frequency ω_k in equation (2.23) the expression (2.9), which interpolates between small and large k behaviours and allows to keep selfconsistent account of the dispersion coefficient α. Selfconsistency is imposed by requiring the value of α used in the SPA response function to coincide with the one computed from f_{xc}^L.

We now briefly discuss the role of two–plasmon excitations in the present context. Starting with a positive value of the dispersion coefficient α in the electron fluid at weak coupling, from the structure of equation (2.23) and the previous results a peak must be expected in $\mathbf{Im} f_{xc}^L(k \to 0, \omega)$ at around twice the plasma frequency. The oscillator strength there must increase with increasing coupling as the plasmon dispersion curve flattens out and the resonance spreads over a wider region of $\mathbf{q}$ space. Thus, as the value of α moves towards zero, an increasingly rapid drop of α is to be expected (see Figure 2.2). The peak may then move towards lower frequencies once α attains negative values, and eventually at large r_s no self–consistent solution is possible.

2.4.3   Numerical results and discussion

In this section we apply the results obtained above to the selfconsistent evaluation of the plasmon dispersion coefficient. Using equation (2.9) in the expressions for the imaginary part of $f_{xc}^L(k, \omega)$ given by equation (2.23), the real part of $f_{xc}^L(k, \omega)$ can be obtained via the Kramers–Kronig relations. One may then compute the dispersion coefficient α from equation (2.26) and hence return to a better estimation of the single–mode frequency in equation (2.9).

TABLE 2.2: Plasmon dispersion coefficient α for bosons as a function of r_s, using different models.

r_s	α_{VS}	α_{Yas}	α_{GK}	α_{K3}
0.1	-0.00644	0.00325	-0.00259	-0.00794
1	-0.0362	0.0140	-0.0177	-0.0445
2	-0.0608	0.0182	-0.0335	-0.0746
3	-0.0822	0.0191	-0.0490	-0.1010
4	-0.1018	0.0181	-0.0643	-0.1253
5	-0.1200	0.0159	-0.0793	-0.1483
6	-0.1372	0.0129	-0.0939	-0.1705
10	-0.199	-0.003	-0.149	-0.258
12	-0.226	-0.012	-0.174	-0.318
15	-0.264	-0.027	-0.211	-
20	-0.322	-0.050	-0.266	-

TABLE 2.3: Plasmon dispersion coefficient α for fermions as a function of r_s, using different models.

r_s	α_{VS}	α_{Yas}	α_{GK}	α_{K3}
0.1	1.99868	2.00737	2.00180	1.99816
1	0.5768	0.6164	0.5884	0.5731
2	0.362	0.426	0.380	0.355
3	0.256	0.340	0.279	0.246
4	0.187	0.288	0.214	0.174
5	0.135	0.252	0.165	0.119
6	0.093	0.223	0.125	0.074
10	-0.026	0.150	0.014	-0.057
12	-0.070	0.124	-0.028	-0.108
15	-0.127	0.092	-0.082	-0.179
20	-0.206	0.049	-0.157	-0.320

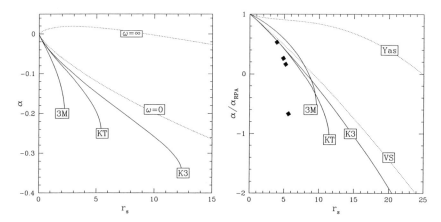

FIGURE 2.13: Selfconsistent plasmon dispersion coefficient α for bosons as a function of r_s for bosons (left panel) and fermions (right panel). In the case of fermions we plotted α/α_{RPA} and included the experimental results by vom Felde *et al* [21] (lozenges). Curves stop where no selfconsistent solution could be obtained.

Evidently, we do have a set of equations to be solved selfconsistently, but with possible alternative choices for the sum rule to be satisfied (equations (2.24) and (2.25)).

The results for the plasmon dispersion coefficient α at selfconsistency are shown in Tables 2.2 and 2.3 and in Figure 2.13. The notation used in these figures is as follows:

(i) K_T denotes a calculation in which $\beta = 1$ and α is determined by satisfying the compressibility sum rule (2.24), thus attributing greater relevance to the low frequency part of the spectrum;

(ii) $3M$ denotes a calculation in which $\beta = 1$ and α is determined by satisfying the third–moment sum rule (2.25), thus attributing greater relevance to the high frequency part of the spectrum;

(iii) $K3$ denotes a calculation in which α and β are determined by satisfying both sum rules.

Our results in Figure 2.13 are also compared with those obtained by neglecting the imaginary part of f_{xc}^L and setting $f_{xc}^L(\omega_{pl})$ equal either to $f_{xc}^L(0)$ or to $f_{xc}^L(\infty)$. The former was done by Vashishta and Singwi [57] and yields an upper bound on both K_T and $K3$, that we denote with VS. The latter was done by Suehiro *et al* [70] and yields an upper bound on both $3M$ and $K3$, that we denote with Yas. Numerical values for both are obtained from the Monte Carlo equation of state. Tables 2.2 and 2.3 report these numerical values for the dispersion coefficient at various values of r_s, together with those

computed from the Gross–Kohn (GK) interpolation formula and with those
that we obtain in our "best" selfconsistent calculation (indicated as α_{K3}). Of
course, the GK and IG values lie between the Yas and the VS results. The
limited accuracy of the fits to Monte Carlo data involved in these evaluations
[71,33] is likely to affect the last significant figure in the numerical values of α
reported in these Tables.

The first part of Figure 2.13 shows the behaviour of α as a function of r_s
for bosons. In all approaches except Yas α is negative throughout the whole
density range. It has recently been shown from sum rule arguments that the
VS result provides an exact upper bound on the plasmon dispersion coefficient
in the degenerate bosonic plasma [55]. Our best result, given by the full curve
in the Figure, correctly lies below this upper bound. The steep drop of α at
$r_s \sim 12$ is due to a breakdown of the SPA as we discuss below in the case of
fermions.

The second part of Figure 2.13 reports the behaviour of α/α_{RPA} as a function of r_s for fermions. All our results show a rapidly increasing drop in the
plasmon dispersion coefficient with increasing r_s and again our best result lies
below the VS result at all values of r_s. This behaviour arises from the mutual
repulsion between the single–plasmon excitation and the two–pair excitations,
with the spectrum of the latter starting within the SPA at twice the minimum
frequency ω_{min} in the single–mode dispersion curve. This repulsion increases
with decreasing energy difference (i.e. at smaller α) and with increasing oscillator strength in the low–frequency part of the two–pair spectrum (again at
smaller α). Therefore, the selfconsistent iterations in α lead to a rapid softening of the plasmon, and eventually no selfconsistent solution can be found.
Evidently the SPA breaks down at a density such that ω_{min} reaches $\omega_{pl}/2$,
where the decay of a plasmon into two modes at $\mathbf{k}_{min}$ and $-\mathbf{k}_{min}$ would become possible.

Finally, Figure 2.13 also reports the measured values of α/α_{RPA} in the
four alkali metals from Na to Cs [21]. It is clear that the plasmon softening
mechanism that we have evaluated still is insufficient to account for these
observations by itself.

2.5   Conclusions

We briefly summarize below the main results that have been obtained on the
kernels $f_{xc}^{L,T}(k,\omega)$ for the homogeneous electron gas at $T = 0$.

- The high frequency limit is given by (1.37–1.38), the leading large–ω
  term is given by (2.7).

- The $(k,\omega) \to (0,0)$ limit is discontinuous, the "static" limit ($\omega = 0$ first)
  is given by the compressibility (longitudinal) and is zero (transverse);

the "dynamic" limit ($k = 0$ first) is given by the elastic bulk and shear
moduli. All of them can be expressed in terms of Landau parameters
(Section 1.5). The combination $f_{xc}^L - 4f_{xc}^T/3$ has the same "static" and
"dynamic" limit.

- An approximate expression including two–pair and in particular two–
  plasmon processes has been obtained (equation 2.3), and its consequences
  examined in detail, finding evidence for:

  - Structure in $\mathbf{Im}\, f_{xc}$ around twice the plasma frequency (Figure 2.6),
    which reflects directly on the dynamic structure factor $S(k, \omega)$ (Fig-
    ure 2.12),

  - Strongly non monotonic behaviour for the real part, at variance
    from previous estimates (Figure 2.7).

- Our results are fitted to analytic expressions (Section 2.2.4), and constitute all the input needed for TD–DFT computations in the long–
  wavelength regime within the Vignale–Kohn scheme, both in the linear
  and in the nonlinear regime (see Section 1.2).

- We presented a self–consistent model for the plasmon dispersion in alkali metals (Section 2.4), which is in good qualitative agreement with
  experimental data.

Part II

Static properties of degenerate quantum plasmas from Diffusion Monte Carlo

Chapter 3

Diffusion Monte Carlo

Diffusion Monte Carlo (DMC) is a well established technique for studying static ground state properties of quantum many–body systems, which allows to obtain exact predictions for a number of quantities [72–74]. Indeed, bosonic ground state energies are subject only to known statistical errors, which can be systematically reduced at the cost of increased computer time. Other properties, such as the structure factor and the momentum distribution, are biased by a trial wave function introduced for computational convenience, but extremely accurate results can be obtained provided the trial function is a good approximation to the true ground state. In the case of fermions additional bias comes from the fixed–node (FN) restriction, which is usually employed to avoid numerical instabilities deriving from changes in the sign of the wavefunction. A number of techniques to improve over FN have been developed [75–77], we shall not discuss them since in the electron gas the FN approximation is not a serious restriction [75,77].

The high accuracy of the DMC results is counterbalanced by a significant computational cost, which restricts the class of systems which can be treated, and by a limited range of applicability. In particular, up to now there is no effective method to study dynamical properties such as those discussed in the previous Chapter. Nevertheless, use of sum rules allows to obtain interesting dynamical informations from the static DMC results, as in the cases of the excitation spectrum displayed in Figure 4.6 and of the asymptotic values of the dynamical exchange–correlation kernels $f_{xc}^{L,T}(k,\omega)$ given in Table 1.1.

We refer to the literature [74,78] for the details of the method. This Chapter summarizes a few essential notions with the aim of characterizing the accuracy of the results to be presented in the following chapters, focussing on the less common features. Chapter 4 presents the results on the charged boson fluid in 3D, and Chapter 5 presents the results on two–dimensional electrons and charged bosons.

3.1   Diffusion algorithm

The DMC procedure basically simulates the Schrödinger equation in imaginary time for a system of N particles as a diffusion equation, yielding a set of configurations* $\{R_i\}$ sampled from the mixed distribution $p(R) = \Phi_0(R)\Psi(R)/\int dR\Phi_0(R)\Psi(R)$. Here Φ_0 is the exact bosonic ground state and Ψ is an explicitly known positive trial function, introduced to improve computational efficiency through importance sampling. The configurations sampled from the mixed distribution $p(R)$ allow the evaluation of mixed expectation values of quantum operators,

$$A_{\mathrm{mix}} = \frac{\langle\Phi_0|A|\Psi\rangle}{\langle\Phi_0|\Psi\rangle} \simeq \frac{1}{M}\sum_{i=1}^{M} A_L(R_i) \qquad (3.1)$$

where $A_L(R) = A\Psi(R)/\Psi(R)$.

The most important example is when A is the Hamiltonian. The quantity to be averaged, called the local energy, reads explicitly

$$H_L(R) = -\frac{\hbar^2}{2m}\left[\sum_{i=1}^{N}\nabla_i^2 \log\Psi(R) + \sum_{i=1}^{N}(\nabla_i \log\Psi(R))^2\right] + \sum_{i<j} v\left(|\mathbf{r}_i - \mathbf{r}_j|\right), (3.2)$$

$v(r) = e^2/r$ being the Coulomb potential. By averaging the local energy one obtains an estimate of the ground state energy, which for bosons is subject only to a known statistical error and to finite–size effects (see Section 3.2).

The statistical uncertainty comes about because of the finite number M of configurations in the sampling. It can be made arbitrarily small for sufficiently long runs provided $A_L(R)$ has a finite variance. Our runs typically consist of $100,000 \div 1,000,000$ Monte Carlo moves after equilibration, a move being an attempt to displace simultaneously all the particles followed by a Metropolis test [74]. The bosonic trial function is of the pair product (Jastrow) form $\Psi(R) = \exp(-\sum_{i<j} u(|\mathbf{r}_i - \mathbf{r}_j|)$, with $u(r)$ the RPA pseudopotential chosen according to Ref. [72]. The fermionic one contains an additional factor to guarantee antisymmetry, which in the present case is a Slater determinant of plane wave states. The FN approximation is then enforced rejecting any move which causes the trial wavefunction to change sign.

The imaginary time evolution in the simulation, i. e. the Monte Carlo move, is generated by a short time approximation of the many–body Green function. The discretization of imaginary time causes a finite time step error, which in principle can be extrapolated out by doing simulations with various time steps. In practice due to our choice of global moves, as opposed to implementations which move only a single particle at a time, we have used a very small time step and verified that the time step effect is not larger than the statistical error for the quantities we measure.

---

*A configuration R is composed by the coordinates $\mathbf{r}_1, \ldots, \mathbf{r}_N$ of all the particles.

3.2   Finite size effects

In our simulations the infinite system is represented by N particles in a cubic box with periodic boundary conditions, and the Coulomb interactions are
treated by the Ewald technique, which splits the potential in a short–range
part to be computed exactly in real space and a long–range part to be represented by k–space sums [72]. The accuracy of the Ewald procedure depends
on the k–space cutoff and on the particular choice for the splitting of the original potential. Remaining finite–size effects on the ground state energy are
significant, and a size extrapolation is needed. This is accomplished assuming
that the size effect is the same in DMC and VMC, and using VMC energies to
determine the finite–size correction as described below. Other quantities have
larger statistical noise and no discernible size effect.

We dealt with these problems slightly differently for different systems, in
the following subsections the various approaches are briefly presented. Section
3.2.4 summarizes an independent evaluation of the 2D and 3D Madelung constants, which have been used to assess the quality of the Coulomb potential
decomposition scheme.

3.2.1   3D bosons

Our simulations on 3D bosons have been performed with 200 particles, using 20
stars of the reciprocal lattice vectors of the simulation box for the k-space sums.
Ewald decomposition was done as specified in Ref. [72], and the calculated
Madelung constant was found to be accurate to 1 part in $2 \cdot 10^5$ (with respect to
the value reported in equation (3.7)). Ewald sums have been used to compute
the pseudopotential $u(r)$ as well, since the wavefunction also has a long-range
tail. This procedure guarantees an accurate representation of the finite system
with periodic boundary conditions, which has much weaker finite-size effects
than a truly finite system, with zero boundary conditions, of comparable size.
Finite-size effects are significant only on the ground state energy, and have
been extrapolated out by assuming a size-dependence of the form $E(\infty) =
E(N) + A(r_s)/N$. The physical motivation of this expression lies in the fact
that, chosen a given particle, $1/N$ of the other ones (in the infinite periodic
system) are periodic images of itself, and their interaction is clearly different
from what should be expected in a truly infinite system. At each density r_s we
performed Variational Monte Carlo (VMC) calculations with various system
sizes, ranging from 27 to 300 particles, and obtained $A(r_s)$ fitting these results,
which in all cases were found to be in good agreement with the assumed $1/N$
size-dependence.

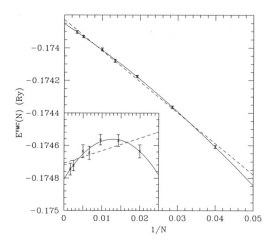

FIGURE 3.1: Size dependence of the VMC energy per particles (dots with error bars) for 2D bosons at $r_s = 10$, as a function of $1/N$. The full curve is the simple fit $E(N) = E(\infty) - A(r_s)/N$, the dashed curve is the improved fit $E(N) = E(\infty) - A(r_s)/N - B(r_s)/N^2$. The inset shows $E(N) + 0.02Ry/N$.

3.2.2   2D bosons

In our 2D simulations we implemented a more refined Ewald decomposition scheme recently proposed by Natoli and Ceperley [79]. The improved decomposition scheme and the lower dimensionality allowed us to limit the k space component to 10 stars only, with a calculated Madelung constant accurate to 1 part in $2 \cdot 10^7$.

In order to reduce the number of particles needed in the simulations we investigated in some detail the size–dependence. In Figure 3.1 we show the data obtained for 2D bosons at $r_s = 10$ and $N =$25, 35, 52, 75, 100, 200 and 300 particles, together with the usual fit $E(N) = E(\infty) - A(r_s)/N$ and an improved fit $E(N) = E(\infty) - A(r_s)/N - B(r_s)/N^2$. The total χ^2 values are 13 and 0.8, respectively. It is evident from these values and from the figure that the $1/N^2$ term must be included in the fitting function to reproduce the MC data. We observed stability of the result for $E(\infty)$ with the improved fit reducing the number of data points used in the fit, and therefore for other densities we used the same expression with only 4 data points.

3.2.3   Fermions

In the case of fermions we adopted the standard procedure of fitting the data to the expression $E(N) = E(\infty) + a/N + b[E_0(N) - E_0(\infty)]$ where $E_0(N)$

is the energy of an ideal Fermi gas as a function of the number of particles. All our simulations were performed at numbers of particles corresponding to "closed shell" configurations for the Slater determinant.

3.2.4   Precise evaluation of Madelung constants

Detailed tests of the code that performs Ewald summations requires knowledge of precise values of the Madelung constant, which is used as a final consistency parameter. In turn, the Madelung constant can be determined as specified by Ceperley [72]. The Coulomb potential is separated in a real–space part

$$v_1(r) = \frac{1}{r\sqrt{\pi}}\Gamma\left(\frac{1}{2}, \alpha r^2\right) \tag{3.3}$$

and a reciprocal space part

$$v_2(k) = \frac{2\pi^{(d-1)/2}}{k^{d-1}}\Gamma\left(\frac{d-1}{2}, \frac{k^2}{4\alpha}\right) \tag{3.4}$$

with

$$v_2(k = 0) = -\frac{2\pi^{(d-1)/2}}{(d-1)\alpha^{(d-1)/2}}. \tag{3.5}$$

We found that for values of α of order unity, in units of the lattice parameter, both series converge quite fast. In fact, a few tens of terms suffice to obtain a fully converged result with 30 significant figures. For cubic lattices in 2 and 3 dimensions we obtained

$$\text{Mad}_{2D} = 3.900\,264\,920\,001\,955\,882\,845\,475\,336\,605 \tag{3.6}$$
$$\text{Mad}_{3D} = 2.837\,297\,479\,480\,619\,476\,665\,917\,104\,608. \tag{3.7}$$

3.3   Extrapolated estimators

In the case of operators not having Φ_0 as an eigenstate (all but the Hamiltonian in the present context), we have calculated the *extrapolated estimator* $A_{\text{ext}} = 2A_{\text{mix}} - A_{\text{var}}$, where $A_{\text{var}} = \langle\Psi|A|\Psi\rangle/\langle\Psi|\Psi\rangle$ is obtained by an independent VMC simulation [73]. The difference between the extrapolated estimator and the true ground state expectation value $\langle\Phi_0|A|\Phi_0\rangle/\langle\Phi_0|\Phi_0\rangle$ is of second order in δ, with $\Psi = \Phi_0 + \delta\Phi$, whereas the bias in the mixed estimator itself is of first order in δ. One can explicitly verify the smallness of δ comparing A_{mix} with A_{var}, or using a more refined wavefunction with additional variationally optimized parameters. Both were checked for a few representative cases, concluding that the remaining δ^2 correction is not larger than the statistical noise.

Quantities which can be routinely computed by this scheme include (i) kinetic and potential energies obtained from the corresponding terms in the local energy, (ii) the pair distribution function

$$g(r) = \frac{1}{Nn} \sum_{i \neq j} \langle \delta(|\mathbf{r}_i - \mathbf{r}_j - \mathbf{r}|) \rangle \ , \tag{3.8}$$

and (iii) the one body density matrix

$$n(r) = \left\langle \frac{\Psi(\mathbf{r}_1, \ldots, \mathbf{r}_i + \mathbf{r}, \ldots, \mathbf{r}_N)}{\Psi(\mathbf{r}_1, \ldots, \mathbf{r}_i, \ldots, \mathbf{r}_N)} \right\rangle \ . \tag{3.9}$$

In these equations $\langle \ldots \rangle$ denotes the average over the configurations $\{R_i\}$ sampled either from $\Phi_0 \Psi$ or from Ψ^2 (in DMC and VMC, respectively). The displacement $\mathbf{r}$ in equation (3.9) can be chosen either by taking fixed increments along a prescribed direction or randomly in the simulation box. The two methods give more accurate results at small and large r respectively, and the former gives a precise value for the curvature of $n(r)$ at $r = 0$. This provides a useful check of internal consistency for the simulation, as the kinetic energy, which is also calculated directly, is proportional to $d^2n(r)/dr^2|_{r=0}$. By Fourier transform of the one–body density matrix one obtains the momentum distribution $n(k)$. This quantity can also be computed directly at reciprocal lattice vectors of the simulation box by

$$n(k) = \left\langle \exp(-i\mathbf{k} \cdot \mathbf{r}) \frac{\Psi(\mathbf{r}_1, \ldots, \mathbf{r}_i + \mathbf{r}, \ldots, \mathbf{r}_N)}{\Psi(\mathbf{r}_1, \ldots, \mathbf{r}_i, \ldots, \mathbf{r}_N)} \right\rangle \ . \tag{3.10}$$

Here $\mathbf{r}$ is a random displacement, and the average $\langle \ldots \rangle$ extends to $\mathbf{r}$. We also average over $\mathbf{k}$ vectors of equal length.

3.4   Static response

The calculation of the static dielectric response $\epsilon(k, 0)$ follows a somewhat different procedure [80,81]. We obtain $\epsilon(k, 0)$ from the static linear density response function $\chi_{\rho\rho}(k)$ via the relationship $1/\epsilon(k, 0) = 1 + v_k \chi_{\rho\rho}(k)$, where $v_k = 4\pi e^2/k^2$ is the Coulomb coupling. In evaluating $\chi_{\rho\rho}(k)$ one perturbs the otherwise homogeneous many–body system with a static external potential

$$v_{ext}(\mathbf{r}) = 2u_\mathbf{k} \cos(\mathbf{k} \cdot \mathbf{r}) \quad . \tag{3.11}$$

This induces a modulation of the density with respect to its mean value n and a shift of the ground state energy per particle, which can be written as

$$E_v = E_0 + \frac{\chi_{\rho\rho}(k)}{n} u_\mathbf{k}^2 + \frac{\chi_{\rho\rho}^{(3)}(\mathbf{k}, \mathbf{k}, -\mathbf{k})}{4n} u_\mathbf{k}^4 + \cdots \tag{3.12}$$

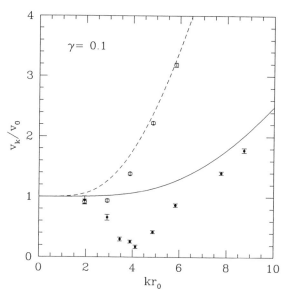

FIGURE 3.2: External field strength u_k for which the optimal value of the variational parameter in the one–body factor is $\gamma = 0.1$, for 3D bosons at $r_s = 10$ (dashed line and circles) and $r_s = 160$ (full line and dots). The curves represent the RPA prediction, while the points are the result of variational minimization.

with $\chi_{\rho\rho}^{(3)}$ the cubic response function. DMC allows one to evaluate E_v for given $\mathbf{k}$ and $u_{\mathbf{k}}$ and by performing simulations at a few coupling strengths $u_{\mathbf{k}}$ one can extract $\chi_{\rho\rho}(k)$ as well as higher–order response functions from the calculated E_v by fitting in powers of $u_{\mathbf{k}}$.

At variance with other simulations performed in this work, for the very time–consuming calculation of $\epsilon(k, 0)$ for 3D bosons we have used only 64 particles. Size effects on the energy difference between the homogeneous and the modulated system have been shown to be very small in Ref. [81].

For the modulated systems the trial function is multiplied by a one–body factor $\Pi_i \exp[-\gamma \cos(\mathbf{k} \cdot \mathbf{r}_i)]$, where the amplitude γ of the density modulation is determined by variational minimization. We show in Fig. 3.2 the external field strength $u_{\mathbf{k}}$ for which $\gamma = 0.1$ is the optimal parameter. For large r_s and $kr_0 \sim 4$ the RPA prediction used in Ref. [81] is far off the result of variational minimization. In principle, DMC should give the exact energy irrespectively of the the trial function. However, a poor choice of γ may give extremely slow convergence to the exact result.

Chapter 4

Charged bosons in 3D

The fluid of point–like spinless charged bosons embedded in a uniform neutralizing background has drawn some attention in the literature as a model in quantum statistical mechanics having a close relationship with the physically more relevant fermionic gas of electrons [82–84]. Its role as a primitive model for the theory of superconductivity was pointed out in the early days by Schafroth [85] and has recently attracted renewed interest in connection with ceramic superconducting materials [86]. The charged boson fluid may also have astrophysical relevance in relation to cores of white dwarf stars consisting of pressure–ionized Helium [87]. Recent interest from an astrophysical point of view was motivated by the study of the fusion of three α–particles in a dense He plasma [88,89], in the framework of the starquake model which has been proposed to explain the frequently observed γ-ray bursts. Injection of deuterium up to a high density into metals like palladium of vanadium was proposed as a promising way to generate a charged Bose gas in laboratory.

On the theoretical side, early evaluations of the ground state energy and the spectrum of elementary excitations in the high–density limit [90–92] were followed by variational calculations of the ground state over a wide range of r_s using Jastrow wave functions [83,84,93,94]. The Hypernetted Chain Approximation (HNC) in the Jastrow theory has been used by Saarela [95] to evaluate the static linear density response function at $T = 0$. Apaja et al [96] have given extensive HNC results on various properties of the boson plasma, including the excitation spectrum. Hore and Frankel [97] had earlier given a full analytic evaluation of the dynamic dielectric function $\epsilon(k, \omega)$ of the fluid at arbitrary temperature within the random phase approximation (RPA). This approximation appears to be particularly restrictive for the fluid of interacting bosons at zero temperature, where the ideal boson gas is fully condensed in the zero–momentum state. The role of correlations beyond the RPA has been explored by Caparica and Hipólito [98] and by Gold [99] within the so–called STLS approximation as proposed earlier for the electron fluid [56]. A more extensive study of exchange and correlation in static local field theories was done in [54].

Quantum Monte Carlo studies of the boson ground state energy [32,84,71,100] have revealed three distinct physical regimes, depending on the dimensionless length parameter $r_s = r_0/a_B$ where r_0 is related to the particle number density n by $r_0 = (4\pi n/3)^{-1/3}$ and a_B is the Bohr radius. The fluid at $r_s \ll 1$ is a weakly coupled gas. The role of correlations increases with decreasing density, leading to a strongly coupled liquid which eventually undergoes Wigner crystallization at $r_s \simeq 160$. This scenario parallels the behavior shown by the fluid of charged fermions. More recently, the static dielectric function $\epsilon(k,0)$ has been directly determined by Monte Carlo methods [81]. Data on the pair distribution function $g(r)$ are also available [101].

This Chapter presents extensive results on the static dielectric function, which has been studied earlier in a very limited range of wavenumbers, and on the momentum distribution, for which the only published results have come from variational Monte Carlo [84]. In addition, structural and thermodynamic data are given over a wide range of density, and we present the first Monte Carlo study of the two–body momentum distribution. In Section 4.6 we discuss the connection of the latter with single particle excitations, and find good agreement between the formal results and the numerical data.

4.1   Ground state energy

The ground state energy at several densities is listed in Table 4.1. These results are in fair agreement with the DMC results of Ceperley and Alder [32], though somewhat higher at small r_s. This small discrepancy is presumably due to differences in the size extrapolation, which is an important correction at high density, being as large as 0.00874 Ry for the system of $N = 200$ particles at $r_s = 1$. There may also be differences in the convergence of the Ewald sums (see Section 3.2). Table 4.1 also compares our data with those reported by Hansen and Mazighi [84], with those obtained through a selfconsistent approach to dielectric screening (STLS) [54,98] and with those obtained within the Hypernetted Chain Approximation (HNC) by Apaja *et al* [96], which are very close to the exact (DMC) result. The variational results of Hansen and Mazighi do not provide an upper bound on the energy in the thermodynamic limit because they have been computed with 128 particles without size extrapolation. In fact they are higher than our data at $N = 128$ but at small r_s lie below our size–extrapolated values.

We have found that both the correct asymptotic behaviours and the DMC data are reproduced by

$$E_g(r_s) = -\left(a_1 r_s^{b_1} + a_2 r_s^{b_2} + (a_3 + a_3' \log(r_s)) r_s^{b_3} + a_4 r_s^{b_4}\right)^c \qquad (4.1)$$

with $a_1 = 7.08556$, $a_2 = 2.20575$, $a_3 = 0.251289$, $a_3' = -0.034817$, $a_4 =$

TABLE 4.1: Ground state energy E_g from DMC simulations compared with the results by Ceperley and Alder [32] (CA), with the variational results by Hansen and Mazighi [84] (HM), with STLS [98,54] and with HNC [96] results at different r_s. All values are in Rydberg per particle, the digits in parentheses represent the error bar in the last decimal place. We also give the kinetic energy $\langle ke \rangle$ and the inverse compressibility $1/nK_T$ as obtained from equation (4.1).

r_s	E_g	CA	HM	STLS	HNC	$\langle ke \rangle$	$1/nK_T$
1	-0.77664(5)	–	-0.7810	-0.771240	-0.77675	0.175217	-0.250743
2	-0.45192(3)	-0.4531(1)	-0.4547	-0.447180	-0.45203	0.095174	-0.148866
5	-0.216420(12)	-0.21663(6)	-0.2170	-0.212895	-0.21675	0.038967	-0.074351
10	-0.121353(5)	-0.12150(3)	-0.1216	-0.118800	-0.12144	0.018247	-0.043501
20	-0.066639(4)	-0.06666(2)	-0.06667	-0.064864	-0.06664	0.007996	-0.024970
50	-0.029276(3)	-0.02927(1)	–	-0.028220	-0.02923	0.002508	-0.011532
100	-0.0154145(13)	-0.015427(4)	-0.01535	-0.014733	-0.01538	0.000997	-0.006257
160	-0.0099046(13)	–	–	–	-0.00988	0.000517	-0.004089

0.009236, $b_1 = 375/56$, $b_2 = 417/56$, $b_3 = 59/7$, $b_4 = 125/14$, and $c = -14/125$.

The parameters in equation (4.1) have been chosen so that for $r_s \to 0$ this formula reproduces the exact limiting expression obtained perturbatively by Brückner, [91]

$$E_g(r_s \to 0) \simeq -\frac{0.8031}{r_s^{3/4}} + 0.0280 + \ldots \tag{4.2}$$

In the opposite limit $r_s \to \infty$ we only impose that the exponents of the two leading terms are -1 and $-3/2$, on the basis of the known asymptotic behaviour for the crystal phase [102,103]

$$E_g(r_s \to \infty) \simeq -\frac{1.79186}{r_s} + \frac{2.65}{r_s^{3/2}} \quad . \tag{4.3}$$

The remaining 4 free parameters have been fitted to the 8 DMC data, obtaining a total χ^2 of 9. The fit is not significantly improved if we do not require c to be a 'simple' rational number.

From the equation of state (4.1) one obtains an unbiased estimator (as opposed to the extrapolated estimator) for the kinetic energy via the virial theorem, $\langle ke \rangle = -d(r_s E_g)/dr_s$. The kinetic energy and the inverse compressibility $1/nK_T = -r_s^4 d((E_g + \langle ke \rangle)/(9r_s^3))/dr_s$ are also listed in Table 4.1.

4.2   Dielectric response

Our results on the static dielectric function $\epsilon(k, 0)$ are shown in Figure 4.1. At low k our data are consistent with the compressibility sum rule

$$\epsilon(k \to 0, 0) \simeq \frac{4\pi e^2}{k^2} n^2 K_T \quad . \tag{4.4}$$

TABLE 4.2: DMC results for the static local field factor $G(k)$ at different r_s. The digits in parentheses represent the error bar in the last decimal place.

kr_0	$r_s = 10$	$r_s = 20$	$r_s = 50$	$r_s = 100$	$r_s = 160$
1.9489	0.29(1)	0.36(3)	0.28(5)	0.38(2)	0.50(4)
2.9233	0.71(1)	0.73(1)	0.75(1)	0.74(1)	0.782(8)
3.5134	-	-	1.005(5)	0.997(9)	0.973(1)
3.8978	1.18(1)	1.21(2)	1.144(5)	1.107(3)	1.080(1)
4.1342	-	-	1.194(7)	1.148(3)	1.119(2)
4.8722	1.54(2)	1.47(1)	1.354(7)	1.210(6)	1.134(6)
5.8467	1.81(7)	1.54(2)	1.26(3)	1.161(7)	1.138(8)
6.8211	2.0(1)	1.53(3)	1.30(5)	1.22(2)	1.17(2)
7.7956	2.4(1)	1.84(8)	1.44(6)	1.53(4)	1.49(2)
8.7700	2.5(2)	2.1(1)	1.8(1)	-	1.75(5)

TABLE 4.3: Values of the fit parameter for the static local field factor $G(k)$, after equation (4.8).

r_s	a	b	c	d	n	$k_0 r_0$	σ
10	0.07185	0.95864	0.020274	0.012929	3.06	3.66636	1.49806
20	0.08061	0.95523	0.017769	0.037192	0.7825	3.46374	1.50438
50	0.08426	0.97756	0.013933	0.048478	0.5435	2.98819	1.77983
100	0.10135	1.02961	0.011078	0.037348	0.7333	3.09635	1.37273
160	0.10878	1.01662	0.009191	0.026138	0.9670	3.31796	1.09357

Given the well–established fact that the compressibility of the fluid of charged bosons is negative at all $r_s > 0$ (see also Table 4.1), equation 4.4 ensures that $1/\epsilon(k, 0)$ goes through a negative minimum before approaching unity as $k \to \infty$. This behaviour implies overscreening of long–wavelength perturbations and does not lead to an instability of the boson plasma, owing to the presence of the rigid background. We remark from Figure 4.1 that the minimum in $1/\epsilon(k, 0)$ becomes deeper with decreasing density and its location approaches $kr_0 \simeq 4$, in approximate correspondence with the first star of reciprocal lattice vectors in the Wigner crystal.

From the DMC data on $\epsilon(k, 0)$ we have obtained the static local field factor $G(k) = -v_k f_{xc}^L(k, \omega = 0)$ for exchange and correlation, defined as in Section 1.4 by

$$\epsilon(k, 0) = 1 - \frac{v_k \chi_{\rho\rho}^0(k)}{1 + G(k) v_k \chi_{\rho\rho}^0(k)} \qquad (4.5)$$

where $\chi_{\rho\rho}^0(k) = -4nm/k^2$ is the static susceptibility of the ideal boson gas.

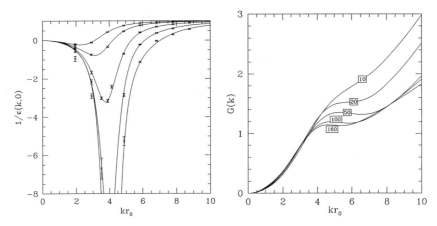

FIGURE 4.1: Left panel: Inverse static dielectric function $1/\epsilon(k,0)$ as a function of kr_0 for $r_s = 10, 20, 50, 100$ and 160. The solid lines are the fit given by equations (4.5) and (4.8). Right panel: Fitted form of the static local field factor $G(k)$ at the same r_s values.

The long wavelength behaviour $G(k \to 0) \simeq -k^2/(4\pi e^2 n^2 K_T)$ is determined by the compressibility sum rule in equation (4.4), while the high–k limit is given by [104]

$$G(k \to \infty) \simeq \frac{4}{3} \frac{\langle ke \rangle}{\omega_{pl}^2} \frac{k^2}{2m} + \frac{16}{5} \frac{\langle ke^2 \rangle}{\omega_{pl}^2} - \frac{16}{9} \frac{\langle ke \rangle^2}{\omega_{pl}^2} + \frac{2}{3}(1 - g(0)) . \qquad (4.6)$$

In equation (4.6) $\langle ke \rangle$ and $\langle ke^2 \rangle$ are the average kinetic energy (see Section 4.1) and the average square kinetic energy, which is determined by the fourth moment of the momentum distribution. Our DMC results for $G(k)$ at various values of r_s are reported in Table 4.2.

For charged fermions in the density range $2 \leq r_s \leq 10$ the DMC results on $G(k)$ [58] show a very smooth crossover from the low–k to the high–k behaviour and can be interpolated with a one–parameter fit to the simple function

$$\frac{G(k)}{(kr_0)^2} = \left[(a - c)^{-n} + \left(\frac{(kr_0)^2}{b} \right)^n \right]^{-1/n} + c \qquad (4.7)$$

with a, b and c given by the asymptotic expressions. Our data for charged bosons are consistent with this picture at $r_s = 10$, but at lower densities $G(k)$ develops a structure around $kr_0 \simeq 4$. Possibly neglecting fine details at very large k of presently unclear relevance, our DMC local field factor is adequately described by simply adding a Gaussian term to the above expression, namely

$$\frac{G(k)}{(kr_0)^2} = \left[(a - c)^{-n} + \left(\frac{(kr_0)^2}{b} \right)^n \right]^{-1/n} + c + d \exp \frac{-(k - k_0)^2 r_0^2}{2\sigma^2} . \qquad (4.8)$$

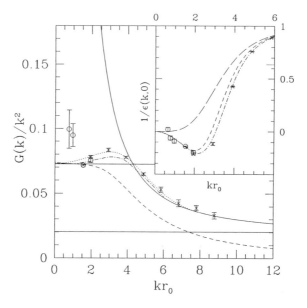

FIGURE 4.2: Static local field factor $G(k)$ over k^2 at $r_s = 10$ (crosses with error bars), compared with the asymptotic expressions (full lines), with the fit (dotted line) and with previous results by other authors. The circles with error bars are the values obtained from the data of Sugiyama *et al* [81], the dashed line gives the VS results from Ref. [54] and the dot–dashed line the variational calculation by Saarela [95]. The corresponding inverse static dielectric function $1/\epsilon(k,0)$ is plotted in the inset, omitting the asymptotic behaviours and the fit and including the RPA curve (long–dashed line).

Fitting this four–parameter expression to the DMC data results in the curves shown in Figure 4.1, with the values of the parameters given in Table 4.3.

We remark that the structure at $kr_0 \simeq 4$ seen in Figure 4.1 for $G(k)$ in the low density regime is not peculiar to the boson fluid. Unpublished data for the polarized electron gas at crystallization density exhibit a similar behavior in both 2 and 3 dimensions [105].

Finally, in Figure 4.2 we compare our data for $G(k)/k^2$ at $r_s = 10$ with its asymptotic behaviours and with the values obtained from the results of Sugiyama *et al* [81]. Their result at $k \simeq 2$ is compatible with ours within statistical errors. The situation at $r_s = 100$ is completely similar. Figure 4.2 also shows the results of approximate theories, obtained from the Jastrow approach [95] and from the selfconsistent VS approach to dielectric screening [54]. The VS curve from Ref. [54] incorporates the correct long–wavelength behaviour, but is not quantitatively correct at higher wave number. The variational calculation by Saarela [95], which incorporates both asymptotic behaviours, is in

TABLE 4.4: Fitted values for the parameters in equation 4.11.

r_s	1	2	5	10	20	50	100	160
n_0	0.827	0.722	0.542	0.359	0.206	0.053	0.0104	0.004
$\langle ke^2 \rangle$	0.745	0.115	0.00863	0.00122	0.000167	$1.24 \cdot 10^{-5}$	$2.05 \cdot 10^{-6}$	$4.72 \cdot 10^{-7}$
a_0	3.8	8.64	12.4	-	-	-	-	-
a_1	1.45	1.75	2.6	-	-	-	-	-
a_2	1.03	1.12	1.3	1.1	1.07	1.5	2.59	2.35
a_3	-	-	0.105	1.3	1.57	0.754	-3.03	-0.77
a_4	0.264	0.531	0.902	0.938	0.693	0.325	0.0901	0.0438
a_5	0.153	0.267	0.422	-0.334	-0.247	0.412	0.245	0.132
a_6	-0.028	-0.0481	-0.0817	0.0615	0.0639	0.165	0.919	0.417

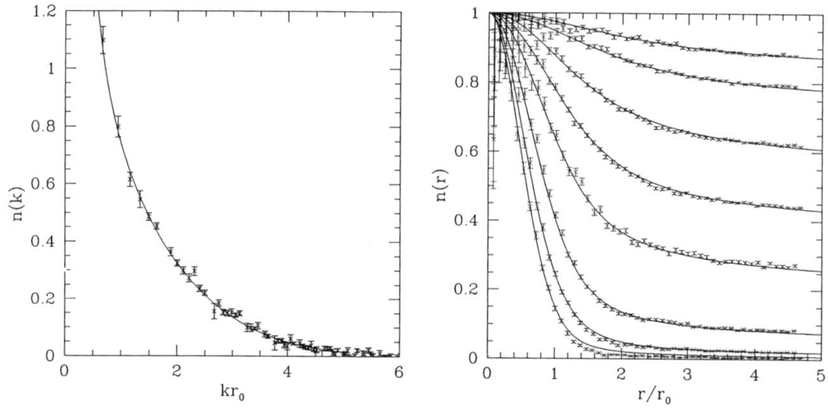

FIGURE 4.3: Left panel: DMC results (crosses with error bars) for the momentum distribution $n(k)$ at $r_s = 100$, compared with the fit given by equation (4.11) (full curve). Right panel: DMC data and fitted expression for the one–body density matrix at $r_s =$1, 2, 5, 10, 20, 50, 100 and 160 (from top to bottom).

remarkably good agreement with our DMC data at all wavenumbers considered.

4.3   Momentum distribution

As mentioned in Section 3.3 we have obtained DMC data for $n(k)$ corresponding to values of $\mathbf{k}$ on the reciprocal lattice of the simulation box and for its Fourier transform $n(r)$ corresponding to values of r within the simulation box. Some of these results are shown in Figure 4.3. We have collected this information in a fitting formula incorporating the known asymptotic behaviours.

At high k the tail of the momentum distribution $n(k)$ is given by the cusp

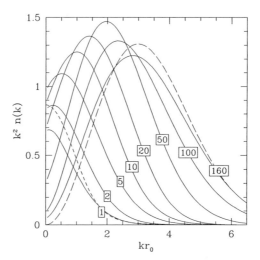

FIGURE 4.4: Momentum distribution at various values of r_s compared with Foldy's results at $r_s = 1$ (short–dashed curve) and with a Maxwell distribution at $r_s = 160$ (dashed curve). We plot $k^2 n(k)$ for greater clarity.

condition [106]

$$n(k \to \infty) \simeq \frac{9r_s^2 g(0)}{(kr_0)^8} \quad . \tag{4.9}$$

In the opposite low–k limit $n(k)$ diverges with a behaviour given by quantum hydrodynamics [107,108]:

$$n(k \to 0) \simeq n_0 \frac{2m\omega_{pl}}{4k^2} + \frac{m^2 n_0 \omega_{pl}^2}{64n\,k} + O(1) \quad . \tag{4.10}$$

Harmonic theory for the crystalline phase predicts a Gaussian momentum distribution, determined by the zero–point phonon energy. We expect that this behaviour should be reflected in that of the liquid close to the transition.

We have therefore chosen the following expression to interpolate the DMC data:

$$n(k) = (2\pi)^3 n_0 n \delta^3(k) + \frac{a_0}{k^2 (k^2 + a_1^2)^3} +$$
$$+ \left(\frac{a_4}{k^2} + \frac{a_5}{k} + a_6 \right) \exp \left\{ -\frac{(k - a_3)^2 - a_3^2}{2a_2^2} \right\} \quad . \tag{4.11}$$

Given the known values of the density and the mean kinetic energy, as well as the known asymptotic behaviours, the remaining parameters are determined by a least–squares fit to the DMC data. We set the parameter a_3 equal to zero when this choice is compatible with the data (as seen from χ^2 values), so that the number of fitted parameters is three or four depending on r_s. The fit was performed on both the k–space points and their Fourier transform, the one–body density matrix. In the fitting procedure we have included only the values of $n(r)$ obtained from random displacements of particles, as mentioned in Section 3.3. In fact, the value obtained by fixed increments is more accurate for small r; however, in this region it basically gives the extrapolated estimator of the kinetic energy, and we prefer to enforce instead the unbiased estimator of $\langle ke \rangle$ of Section 4.1. In order to give the same weight to real space and to reciprocal space the function we minimized was the sum of the two reduced χ^2 values, which were typicallly found to be in the range from 0.3 to 2.5.

Figure 4.3 shows the k–space data and the fitted curve for $r_s = 100$ as well as the corresponding $n(r)$ at various values of r_s. The $r \to \infty$ limiting value of $n(r)$ gives the condensate fraction.

Figure 4.4 shows the resulting expressions for $k^2 n(k)$ at various values of r_s. At $r_s = 1$ the Bogolubov curve [90] is almost correct, as could be expected from the fact that the condensate fraction is above 80%. This is consistent with the usual picture of $r_s \to 0$ as the 'uniform' (or RPA) limit. With growing r_s the shape of the curves changes qualitatively, with exchange effects playing a minor role at low density. At the crystallization density ($r_s = 160$) less than 1% of the particles are in states with occupation number larger than 1 and the resulting momentum distribution is similar to the Maxwell–Boltzmann distribution taken at a fictitious temperature given by $\langle ke \rangle = \frac{3}{2} k_B T$.

4.4   Structure factor

The static structure factor $S(k)$ has been obtained by Fourier transforming the DMC data on the pair distribution function $g(r)$ as given by the simulation for $r < L/2$, with L the side of the simulation box, after extending the data for $r > L/2$ with an oscillatory, decreasing tail of the form $ar^{-b} \sin(c(r - r_0))$. The parameters a, b, c and r_0 are fitted to the higher–r part of the data. This procedure gives reliable results for intermediate and high k, but does not reproduce very well the $k \to 0$ limit since it does not contain any normalization condition. The resulting $S(k)$ is then smoothly matched to the correct asymptotic behaviour for $k \to 0$, the result being shown in Figure 4.5. A strong peak at $kr_0 \simeq 4$ develops as the crystallization density is approached.

The statistical errors in $S(k)$ are estimated to be less than 1%, smaller than the systematic errors coming from the use of extrapolated estimators (see Section 3.3) and from the tail added to $g(r)$. The final precision can be estimated to be of a few percent.

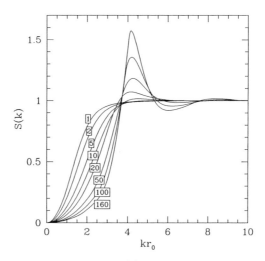

FIGURE 4.5: Static structure factor $S(k)$ as a function of kr_0 for $r_s = 1, 2, 5, 10,$ 20, 50, 100 and 160.

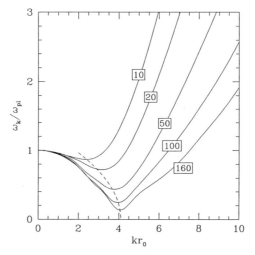

FIGURE 4.6: Upper bound on the plasmon dispersion curve as derived from the static susceptibility and the static structure factor, for $r_s = 10, 20, 50, 100$ and 160. The dashed line shows the position of the minimum in the various curves.

4.5   Plasmon dispersion

The so–called sum–rules approach allows to extract dynamical information from knowledge of static properties. Starting from the Feynmann estimate $\omega_k = nk^2/mS(k)$ of the collective excitation energy ω_k, very useful informations have been obtained on liquid ^{4}He [109]. The availability of quantitative results for static properties from Monte Carlo simulations gives new relevance to this method. This section addresses some sum rules on the density excitation spectrum of the charged boson plasma, whereas the next section discusses single particle excitations.

The sum rules presented in Chapter 1 for the case of fermions apply also for bosons, with the same proof. In terms of the moments of the density response function

$$m_n(k) = -\frac{2}{\pi} \int_0^\infty \mathbf{Im}\, \chi_{\rho\rho}(k,\omega)\omega^n d\omega \qquad (4.12)$$

they state [30]

$$m_{-1}(k) = -\chi_{\rho\rho}(k,0) = \frac{1}{v_k} - \frac{1}{n^2 K_T v_k^2} + O(k^6) \quad , \qquad (4.13)$$

$$m_0(k) = 2nS(k) \quad , \qquad (4.14)$$

$$m_1(k) = \frac{nk^2}{m} = \frac{\omega_{pl}^2}{v_k} \quad , \qquad (4.15)$$

$$m_3(k) = \frac{nk^2}{m}\left(\omega_{pl}^2 + 4\varepsilon_k\left(\langle ke\rangle - \frac{2}{15}\langle pe\rangle\right)\right) + O(k^6) \quad . \qquad (4.16)$$

The results for m_{-1}, m_1 and m_3 show that the long wavelength limit of $\omega^{-1}v_k\theta(\omega)\mathbf{Im}\,\chi_{\rho\rho}(\omega)$ is a probability distribution with average ω_{pl} and zero variance, i.e. a delta function. It follows that $S(k\to 0) \simeq k^2/2\omega_{pl}$.

At finite k the above expressions furnish uppers bounds on the plasmon dispersion curve, or more precisely on the energy $\omega_k^{\min}$ of the lowest eigenstate excited by the operator $\rho_\mathbf{k} = \sum_\mathbf{q} a_\mathbf{q}^\dagger a_{\mathbf{q}+\mathbf{k}}$. The moments give various averages of the spectral strength, which are obviously higher than the lowest attained value $\omega_k^{\min}$. It follows that

$$\omega_k^{\min} \le \sqrt{\frac{m_1(k)}{m_{-1}(k)}} = \omega_{pl} + \frac{k^2}{2m\omega_{pl}nK_T} + O(k^4) . \qquad (4.17)$$

The compressibility K_T of the boson plasma is negative at any r_s (see Section 4.1), the above result proves that the long–wavelength plasmon dispersion is also negative [55]. Indeed, in the case of fermions the plasmon dispersion turns out to be negative at large r_s, when correlation dominates over statistics (see Section 2.4 and Ref. [110]).

The more stringent bound

$$\omega_k^{\min} \le \frac{m_0(k)}{m_{-1}(k)} = -\frac{2nS(k)}{\chi_{\rho\rho}(k,0)} \qquad (4.18)$$

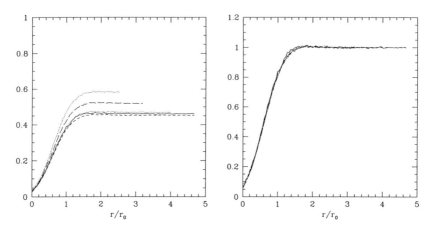

FIGURE 4.7: Left panel: The function $f(r) = n_0(1 + F_1(r))$ at $r_s = 10$ as a function of r/r_0, where the mean interparticle distance r_0 is given by $(4\pi n/3)^{-1/3}$. VMC results with $N = 32$, 64, 125 and 200 are shown, from top to bottom. The full curve is a DMC calculation at $N = 200$. Right panel: The same data after division by their large-r limit.

gives the curves displayed in Figure 4.6 at various values of r_s. In the case of liquid ^{4}He direct comparison with experimental data [109] shows that, while the Feynman result is about a factor of 2 higher than the roton minimum, the bound given by (4.18) is only 30% too high. However, as for ^{4}He one may expect that the approach of the upper bound to the single particle recoil frequency at high k is qualitatively incorrect.

We also notice from Figure 4.6 that with increasing r_s the minimum in the upper bound decreases down to very low frequency in approximate correspondence with the first Brillouin zone edge of the Wigner crystal.

4.6 Two–body momentum distribution and single particle excitations at long wavelength

This section presents a study of the two–body momentum distribution, which allows to extract information on the single particle excitation spectrum. Indeed, the average energy $\varepsilon_P(k)$ of the single–particle excited state $|P_{\mathbf{k}}\rangle = a_{\mathbf{k}}|0\rangle$ is given by

$$\varepsilon_P(k) = \frac{\langle P_{\mathbf{k}}|H - \mu N|P_{\mathbf{k}}\rangle}{\langle P_{\mathbf{k}}|P_{\mathbf{k}}\rangle} = \frac{\langle a_{\mathbf{k}}^\dagger[H - \mu N, a_{\mathbf{k}}]\rangle}{\langle a_{\mathbf{k}}^\dagger a_{\mathbf{k}}\rangle} = \mu - \frac{k^2}{2m} - \frac{1}{n(k)}\sum_{\mathbf{q}\neq 0} n(\mathbf{k}, \mathbf{q})v(q),$$

$$(4.19)$$

where $n(\mathbf{k}, \mathbf{q}) = \langle a_{\mathbf{k}}^{\dagger} \rho_{\mathbf{q}} a_{\mathbf{k}-\mathbf{q}} \rangle$ is the two–body momentum distribution. The $\mathbf{q} = \mathbf{k}$ term in equation (4.19) is singular because the $a_{\mathbf{q}-\mathbf{k}}$ operator in $n(\mathbf{k}, \mathbf{q})$ acts on the condensate, therefore we compute its $k \to 0$ limit separately. The terms with $\mathbf{q} \neq 0, \mathbf{k}$ are regular at small k, yielding

$$\frac{1}{VNn_0} \sum_{\mathbf{q} \neq 0} n(0, \mathbf{q}) v(q) = \int \frac{d^3 \mathbf{q}}{(2\pi)^3} v(q) F_1(q) \qquad (4.20)$$

where $F_1(q)$ is defined by

$$n(0, k) = \sqrt{Nn_0} \langle \rho_{\mathbf{k}} a_{-\mathbf{k}} \rangle = Nn_0 F_1(k). \qquad (4.21)$$

The Fourier transform of $F_1(q)$ was first introduced in the case of ^{4}He by Ristig and Clark [111] to characterize the long–range behaviour of the two–body density matrix

$$\lim_{|\mathbf{r}_1 - \mathbf{r}'_1| \to \infty} \rho_2(\mathbf{r}_1, \mathbf{r}_2; \mathbf{r}'_1, \mathbf{r}_2) = n_0 n^2 \left[1 + F_1(|\mathbf{r}_1 - \mathbf{r}_2|) + F_1(|\mathbf{r}'_1 - \mathbf{r}_2|) \right] \quad (4.22)$$

or equivalently

$\lim_{\mathbf{r}'_1 \to \infty} \rho_2(\mathbf{r}_1, \mathbf{r}_2; \mathbf{r}'_1, \mathbf{r}_2) = n_0 n^2 [1 + F_1(|\mathbf{r}_1 - \mathbf{r}_2|)]$. It follows that $F_1(r = \infty)$ vanishes, and the normalization condition on ρ_2 then yields $F_1(k = 0) = -1/2$. This allows to compute the long–wavelength limit of the $\mathbf{q} = \mathbf{k}$ term in equation (4.19),

$$\lim_{k \to 0} \frac{1}{V} \frac{1}{n(k)} n(\mathbf{k}, \mathbf{k}) v(k) = \lim_{k \to 0} \frac{1}{V} \frac{4\varepsilon_k}{n_0 \omega_{pl}} Nn_0 F_1(k) v_k = -\omega_{pl} \qquad (4.23)$$

where equation (4.10) has been used. This proves that

$$\lim_{k \to 0} \varepsilon_P(k) = \omega_{pl} + \mu - \int \frac{d^3 \mathbf{q}}{(2\pi)^3} v(q) F_1(q) \qquad (4.24)$$

and therefore the relation $\lim_{k \to 0} \varepsilon_P(k) = \omega_{pl}$ is equivalent to the relation of Hugenholtz and Pines [112],

$$\mu = \int \frac{d^3 \mathbf{q}}{(2\pi)^3} v(q) F_1(q) \quad . \qquad (4.25)$$

We now proceed to present a Monte Carlo study of the function $F_1(k)$, including a numerical test of the mentioned sum rules. We have performed both VMC and DMC simulations at various numbers of particles, ranging from 32 to 200. A separate analysis has been made using real space and reciprocal space estimators. The real–space estimator is

$$f(|\mathbf{r}_1 - \mathbf{r}_2|) = \frac{1}{V} \frac{1}{n^2} \int_V d^3 \mathbf{r}'_1 \rho_2(\mathbf{r}_1, \mathbf{r}_2; \mathbf{r}'_1, \mathbf{r}_2) = \frac{1}{V} \int_V d^3 \mathbf{r}'_1 \left\langle \frac{\Psi(\mathbf{r}'_1, \mathbf{r}_2, \ldots)}{\Psi(\mathbf{r}_1, \mathbf{r}_2, \ldots)} \right\rangle$$
$$(4.26)$$

TABLE 4.5: Computed values of I_1 and I_2 defined in equations (4.27) and (4.28) from real-space estimators. The digits in parentheses represent the statistical error bar in the last decimal place. No size extrapolation has been made. Energies and frequencies are expressed in Rydberg.

r_s	N	I_1	I_2	μ	ω_{pl}
10	64	-0.52(2)	-0.162(3)	-0.1557	0.1095
10	125	-0.46(4)	-0.153(4)	-0.1557	0.1095
10	200	-0.48(2)	-0.153(2)	-0.1557	0.1095
10(DMC)	200	-0.48(2)	-0.151(2)	-0.1557	0.1095
1	200	-0.46(4)	-0.93(3)	-0.977	3.46
0.1	200	-0.45(4)	-5.2(1)	-5.62	109.5

TABLE 4.6: Computed values of I_2 defined in equation (4.28) from reciprocal-space estimators. Energies and frequencies are expressed in Rydberg.

r_s	N	I_2	μ	ω_{pl}
10	64	-0.1546	-0.1557	0.1095
10	125	-0.1604	-0.1557	0.1095
10	200	-0.1565	-0.1557	0.1095
10(DMC)	200	-0.1558	-0.1557	0.1095
1	200	-0.917	-0.977	3.46

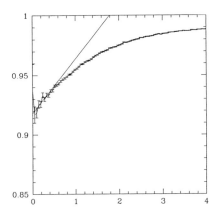

FIGURE 4.8: $F_1(r)$ at $r_s = 0.1$ as a function of r/r_0 from VMC with $N = 200$. Also shown is the asymptotic behaviour given by the cusp condition in equation (4.29).

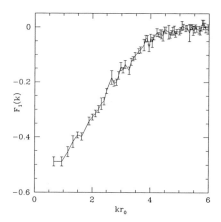

FIGURE 4.9: The function $F_1(k)$ vs. kr_0 at $r_s = 10$ from DMC with $N = 200$.

which equals $n_0(1 + F_1(|\mathbf{r}_1 - \mathbf{r}_2|))$ in the thermodynamic limit. Figure 4.7 shows the function $f(r)$ at $r_s = 10$ for $N = 32, 64, 125$ and 200. The size effect which is apparent in the data in Figure 4.7 arises mainly from the regions where $\mathbf{r}'_1 \simeq \mathbf{r}_2$ or $\mathbf{r}'_1 \simeq \mathbf{r}_1$, which are of relative importance $1/N$. A precise estimate of the thermodynamic limit can be obtained by dividing each curve by its value at large r, as is shown in the left panel of Figure 4.7.

These curves are then integrated to determine the values of the integrals

$$I_1 = n \int_V F_1(r)d^3\mathbf{r} = F_1(k = 0) \tag{4.27}$$

and

$$I_2 = n \int_V F_1(r)v(r)d^3\mathbf{r} = \int \frac{d^3\mathbf{k}}{(2\pi)^3}F_1(k)v(k) \quad . \tag{4.28}$$

According to the previous discussion these ought to be equal to $-1/2$ and to the chemical potential μ, respectively. The results are shown in Table 4.5 together with the chemical potential as obtained from the interpolation between the DMC data presented in Section 4.1. The table also reports the plasma frequency, showing that it is much larger than the remaining differences between μ and I_2 and thus confirming that $\varepsilon_P(k)$ goes to ω_{pl} for $k \to 0$ as shown in equation (4.24).

We also remark that good agreement has been found with the appropriate cusp condition

$$\left.\frac{d}{dr}F_1(r)\right|_{r=0} = \frac{1}{2a_B}F_1(r = 0) \quad , \tag{4.29}$$

as is shown in Figure 4.8 for $r_s = 0.1$. As usual, the cusp condition is relevant only at large densities, where $F_1(r = 0) \neq 0$.

As in the case of the momentum distribution, an independent estimate can be obtained from the reciprocal–space estimator

$$f(k) = \frac{1}{n^2} \int_V \frac{d^3\mathbf{r}'_1}{V} e^{i\mathbf{k}\cdot(\mathbf{r}_1-\mathbf{r}_2)} \rho_2(\mathbf{r}_1, \mathbf{r}_2; \mathbf{r}'_1, \mathbf{r}_2) = \int_V \frac{d^3\mathbf{r}'_1}{V} \left\langle e^{i\mathbf{k}\cdot(\mathbf{r}_1-\mathbf{r}_2)} \frac{\Psi(\mathbf{r}'_1, \mathbf{r}_2, \ldots)}{\Psi(\mathbf{r}_1, \mathbf{r}_2, \ldots)} \right\rangle$$
(4.30)

which equals $n_0(F_1(k) + N\delta_{\mathbf{k}0})$ in the thermodynamic limit. The condensate fraction is obtained as a byproduct from the relation $f(k = 0) = Nn_0$.

Figure 4.9 shows the resulting DMC estimate of $F_1(k)$, which is clearly consistent with the theoretical limit $F_1(k \to 0) = -1/2$. Integration over k leads to significant averaging out of the errors and to a very precise estimate of the integral I_2 at intermediate density, reported in Table 4.6. This kind of computation is more difficult at high densities because the momentum distribution is very narrow and only a few sampling wave vectors fall into the region where it is significantly different from zero. Strong size effects have also been observed in this regime, implying that the result at $r_s = 1$ in Table 4.6 is less reliable than those at $r_s = 10$. For the same reasons, it has not been possible to obtain reliable results at $r_s = 0.1$.

4.7 Conclusions

We have presented the first systematic study of the properties of the charged boson fluid at $T = 0$ by Diffusion Monte Carlo. We have given a simple analytical expression for the internal energy of the fluid as a function of coupling strength r_s, from which reliable results for the equation of state can be obtained in the whole fluid density range $0 \leq r_s \leq 160$. Our results for the static density response function cover the relevant region of wavenumbers and show significant changes in the response as the fluid approaches crystallization density. An analytical interpolation of these data has also been given. Similar features emerge from the data on the structure factor. The shape of the momentum distribution changes significantly in the evolution of the system from the uniform limit ($r_s < 1$) to the strongly–coupled fluid regime ($r_s \simeq 160$), paralleling the decrease of the condensate fraction from 83% to less than 1%. These data have also been summarized in an interpolation formula consistent with the known asymptotic behaviours. As an application we have presented a rigorous upper bound on the plasmon dispersion curve, which shows the presence of a deep minimum in the strong coupling regime. We finally presented a Monte Carlo study of the two–body momentum distribution, which can be related to the dispersion of single particle excitations, finding good agreement with theoretical expectations.

Chapter 5

Charged bosons and fermions in 2D

Two dimensional electron gas (2DEG) systems confined in semiconductor space charge layers (*e.g.*, Si inversion layers, GaAs heterojunctions, and quantum wells) have provided for the last twenty years an ideal laboratory for studying various electron–electron interaction effects under almost ideal 2D jellium conditions. The zero temperature phase diagram has attracted a lot of attention since Wigner [113] pointed out that at low density electrons would crystallize to minimize the potential energy. A stable spin polarized phase at intermediate density was predicted by Hartree–Fock (HF) calculations, as a consequence of competition between kinetic and exchange energy. The most recent Diffusion Monte Carlo (DMC) simulations by Rapisarda and Senatore (RS) [114] confirm this picture and show that the 2DEG is paramagnetic up to $r_s = 20$, then undergoes a ferromagnetic transition and eventually crystallizes at $r_s = 34$. As usual, the coupling strength r_s is defined in terms of the areal density by $n_{2D} = 1/\pi(r_s a_B)^2$, a_B being the Bohr radius. The only QMC simulation data for the momentum distribution come from Variational Monte Carlo (VMC) calculations [115] and are rather incomplete. We present extensive DMC results for the both momentum distribution $n(k)$ and the one body density matrix $n(r)$, and independently recover some of the results of RS.

The analogous bosonic system has been considerably less studied in the literature. It was first discussed by May [116] in relation to Meissner effect in thin films of superconducting materials. Interest has been growing in the last years, mainly due to possible analogies with cuprate superconductors and to bosonic excitations in heterostructures. Recent theoretical studies using static local field theories [99,117,118] and correlated basis function methods [96,119] give for the liquid phase a picture similar to the three dimensional case (see Chapter 4). To the best of our knowledge no quantitative estimate has yet been made on the appearance of a solid phase and, in general, no MC results

TABLE 5.1: Variational Monte Carlo energies for 2D bosons at $r_s =2, 10, 20, 40$ and 75 for various numbers of particles N and in the thermodynamic limit. Size extrapolation was done as explained in Sectino 3.2.2. We also give the total χ^2 values for the size–extrapolation and DMC results with 52 particles and in the thermodynamic limit.

N	$r_s = 2$	$r_s = 10$	$r_s = 20$	$r_s = 40$	$r_s = 75$
20	–	-0.174864(6)	–	–	–
25	–	-0.174623(5)	-0.093189(5)	–	–
35	-0.67777(4)	-0.174369(5)	-0.093104(5)	-0.048805(2)	-0.0268518(8)
52	-0.67557(3)	-0.174176(5)	-0.093035(4)	-0.048783(2)	-0.0268420(7)
75	-0.67427(6)	-0.174064(8)	-0.092992(3)	-0.048768(2)	-0.0268367(9)
100	-0.67367(3)	-0.174006(6)	-0.092971(3)	-0.048761(2)	-0.0268332(7)
200	–	-0.173927(4)	–	–	–
300	–	-0.173906(5)	–	–	–
VMC_∞	-0.67192(6)	-0.173852(3)	-0.092903(3)	-0.048737(2)	-0.026825(1)
χ^2	0.7	0.8	0.3	0.1	0.2
DMC_{52}	-0.6776(1)	-0.17513(5)	-0.093519(6)	-0.049032(7)	-0.026983(6)
DMC_∞	-0.6740(2)	-0.17480(5)	-0.093387(8)	-0.048986(8)	-0.026965(6)

are present in the literature for this system. We present below DMC results for the equation of state, the pair correlation function and the momentum distribution and compare with existing theories.

5.1   2D bosons

5.1.1   Ground state energy

Tables 5.1 and 5.2 report our simulation results at different number of particles. Size extrapolation was done as specified in Section 3.2.2. The resulting DMC energies in the thermodynamic limit are compared in Table 5.3 with the results obtained with STLS by Gold [99], with a parametrized wave function approach by Sim, Tao and Wu [119] and with HNC by Apaja et al [96]. As in the case of 3D bosons (Section 4.1) the agreement between the HNC and DMC results is particularly good.

Our DMC results can be accurately reproduced with a fitting function analogous to the one adopted for 3D bosons in Section 4.1. We used

$$E(r_s) = - \left[a_0 r_s^{b_0} + a_1 r_s^{b_1} + a_2 r_s^{b_2} + a_3 r_s^{b_3} \right]^{-c} Ry \qquad (5.1)$$

where a_0 and b_0 are fixed by the small–r_s behaviour [96]

$$E(r_s \to 0) \simeq - \frac{1.29355}{r_s^{2/3}} Ry , \qquad (5.2)$$

TABLE 5.2: Analogous of Table 5.1, for $r_s = 5$. The difference in the number of particles used is fortuitous.

N	$r_s = 5$
26	-0.31959(2)
42	-0.31868(2)
52	-0.31842(2)
58	-0.31831(2)
114	-0.31787(2)
VMC_∞	-0.317456(7)
χ^2	0.001
DMC_{52}	-0.31999(5)
DMC_∞	-0.31903(5)

N	$r_s = 1$
35	-1.15915(6)
52	-1.15301(6)
75	-1.14950(8)
100	-1.14759(8)
180	-1.14542(8)
VMC_∞	-1.14278(9)
χ^2	1.3
DMC_{52}	-1.1550(3)
DMC_∞	-1.1448(5)

TABLE 5.3: Ground state energy for bosons from VMC and DMC, extrapolated to the bulk limit and compared with the HNC computations by Apaja *et al* from Ref. [96], the parametrized wave function computations by Sim Tao and Wu (STW) [119] and the STLS results by Gold [99].

r_s	$E^{(DMC)}$	$E^{(VMC)}$	HNC	STW	STLS	$\langle ke \rangle$	$1/nK_T$
1	-1.1448(5)	-1.14278(9)	-1.1458	-1.1062	–	0.2903	-0.5731
2	-0.6740(2)	-0.67192(6)	-0.6740	-0.6631	-0.6484	0.1442	-0.3582
5	-0.31903(5)	-0.317456(6)	-0.3185	-0.3133	-0.3078	0.04896	-0.187
10	-0.17480(5)	-0.17385(3)	-0.1741	-0.16685	-0.1724	0.01961	-0.1097
20	-0.093387(8)	-0.092903(3)	-0.0928	-0.086024	-0.0959	0.007533	-0.06177
40	-0.048986(8)	-0.048737(2)	–	–	–	0.00286	-0.03359
75	-0.026965(6)	-0.026825(1)	–	–	–	0.001189	-0.01892

b_1 is fixed by requiring a constant subleading term for $r_s \to 0$, b_2 and b_3 by requiring a $ar_s^{-1} + br_s^{-3/2}$ behaviour for $r_s \to \infty$. The final values of the parameters are $c = 7/40$, $a_0 = 0.2297$, $a_1 = 0.161$, $a_2 = 0.0594$, $a_3 = 0.01017$, $b_0 = 80/21$, $b_1 = 94/21$, $b_2 = 73/14$, and $b3 = 40/7$. The total χ^2 for the fit with 4 parameters and 7 data points is 4.5.

Figure 5.1 compares our results with the previous DMC results by Rapis-arda and Senatore [114] on 2D fermions and on the 2D Wigner crystal, statistics being irrelevant in the solid phase. In two dimensions bosons crystallize at $r_s \simeq 60$, and fermions at $r_s \simeq 34$. The difference in critical density is analogous to the difference obtained in the 3D case, where bosons crystallize at $r_s = 160$ and fermions at $r_s = 100$ [32].

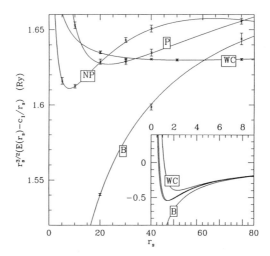

FIGURE 5.1: Ground state energy of 2D triangular Wigner crystal (WC), bosons (B), non polarized (NP) and polarized (P) fermions as a function of r_s. Wigner crystal and fermion data are from RS. On the purpose of clarity we plotted $r_s^{3/2}(E(r_s) - c_1/r_s)$, with $c_1 = -2.2122$, while the inset shows the corresponding $E_g(r_s)$ curves. Points with error bars are size–extrapolated DMC results, continuous curves are analytical fits given by equation (5.1) for bosons, the remaining as described in RS.

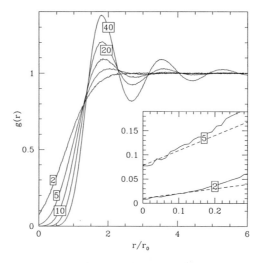

FIGURE 5.2: Pair correlation function $g(r)$ for bosons at $r_s = 1, 2, 5, 10, 20$ and 40. The inset compares the small–r data at $r_s = 1, 2$ and 5 with the cusp condition $g'(0) = 2g(0)/a_B$ [120].

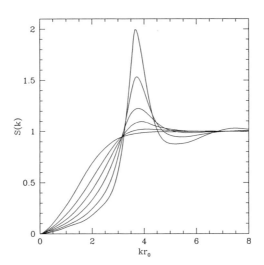

FIGURE 5.3: Static structure factor of 2D bosons at $r_s =$2, 5, 10, 20, 40 and 75. Higher r_s values correspond to a higher peak and to a lower value at small k.

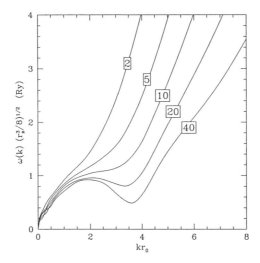

FIGURE 5.4: Feynman upper bound on the plasmon dispersion for 2D bosons at $r_s =$2, 5, 10, 20 and 40.

5.1.2 Structure factor

Figure 5.2 contains the static pair correlation function $g(r)$ for 2D bosons at different r_s values. The static stucture factor $S(k)$ reported in Figure 5.3 was obtained as in the 3D case (see Section 4.4). A sharp peak develops in the strongly correlated ($r_s = 20$ and 40) and metastable ($r_s = 75$) liquid phase in approximate correspondence with the first reciprocal lattice wavevector of the 2D Wigner crystal, $k_0 r_0 = \sqrt{2\pi\sqrt{3}} \simeq 3.3$. The peak height is around 1.6 at crystallization density, as in other quantum fluids (For 3D bosons see Figure 4.5; for 2D fermions see below; for ^{4}He see Ref. [121]). The difference with the classical value $\sim 2.5 \div 3$ is analogous to the known difference in Lindeman ratio at melting.

At the same time, the Feynman expression for the collective excitations $\omega_k = k^2/2mS(k)$ displayed in Figure 5.4 exhibits a pronounced minimum around the same wavelength. We remark that the present result is an upper bound on the excitation spectrum in the sense of Section 4.5, but that it is a worse (i.e. higher) bound than the one discussed for 3D bosons.

5.1.3 Momentum distribution

Size effects on the large–r tail of the one body density matrix $n(r)$ are quite large (see Figure 5.5). In the case of 2D bosons with $\ln r$ interactions Magro and Ceperley [122] have shown that the strongest size effect on $n(r)$ derives from application of periodic boundary conditions, which in the computation of $n(r)$ mean that one evaluates the change in the trial wavefunction resulting from a displacement of a particle and *all its periodic images*. This effect can be eliminated introducing in equations (3.9–3.10) the factor

$$\exp\left\{\frac{1}{2}\left[\sum_{\mathbf{q}\neq 0} u_{\mathbf{q}}(e^{i\mathbf{q}\cdot\mathbf{r}} - 1) - [u(\mathbf{r}) - u(0)]\right]\right\} \qquad (5.3)$$

where $\mathbf{q}$ runs over the reciprocal lattice of the simulation box and $u_{\mathbf{q}}$ is the Fourier transform of the two–body pseudopotential, as defined in Section 3.1. This correction need not be included in the evaluation of $n(k)$ since it corresponds to the introduction of a non–periodic term, which increases finite–size effects on the k–space results.

Figure 5.5 compares our VMC results at $N = 52$ and at $N = 100$, with and without the correction described in equation (5.3), and shows that size–effects are virtually eliminated – within statistical uncertainties – by the above correction. Furthermore, Figure 5.6 shows that the Fourier transform of a fit to the reciprocal–space data (done as explained below, but without using any $n(r)$ data) is in good agreement with the real–space data if and only if the correction (5.3) is properly included.

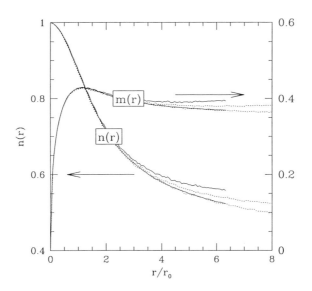

FIGURE 5.5: One–body density matrix $n(r)$ computed with $N = 52$ (full curves) and $N = 100$ (dotted curves) particles for 2D bosons at $r_s = 2$ with the standard estimator (higher couple of curves) and with the corrected estimator from equation (5.3) (lower couple of curves). The curves labeled "$m(r)$" are the same data rescaled according to equation (5.5) [$m(r) = n(r)/(1 + 0.73967\sqrt{r_s/r})$]. The large–$r$ limit of $m(r)$ as obtained with the correction (5.3) gives a reliable estimate of n_0 (scale on the right).

The stronger size effects observed in 2D with respect to the 3D case can be attributed to the slower convergence of $n_0(r)$ to its asymptotic value. In fact, in 3D $n(k \to 0) \propto k^{-2}$ (equation (4.10)) implies $n(r \to \infty) - n_0 \propto r^{-1}$. The divergence of the two–dimensional momentum distribution at small k can determined along the lines of Ref. [107], one gets

$$n(k \to 0) \simeq \frac{n_0}{4S(k)} \simeq \frac{n_0\sqrt{r_s/2}}{(kr_0)^{3/2}} \qquad (5.4)$$

where n_0 is the condensate fraction, $r_0 = r_s a_B$ and the long–wavelength structure factor has the exact limiting behaviour $S(k) \simeq (kr_0)^{3/2}/\sqrt{8r_s}$. In real space this gives

$$n(r \to \infty) - n_0 \simeq n_0 \frac{\Gamma(1/4)}{4\Gamma(3/4)} \sqrt{\frac{r_s r_0}{r}} \simeq 0.73967 \cdot n_0 \sqrt{\frac{r_s r_0}{r}} . \qquad (5.5)$$

The above result allows to obtain reliable estimates of n_0 directly from the

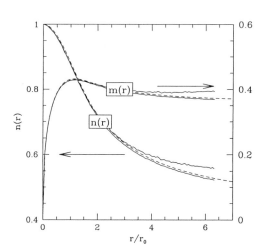

FIGURE 5.6: Same as Figure 5.5, but only the data with $N = 52$ are shown. The dashed curve is the fit to the reciprocal space data obtained neglecting the correction (5.3).

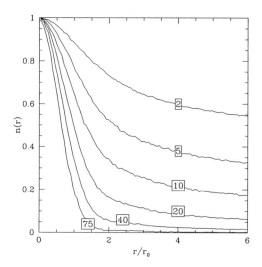

FIGURE 5.7: One body density matrix $n(r)$ for bosons at $r_s = 2, 5, 10, 20, 40$ and 75.

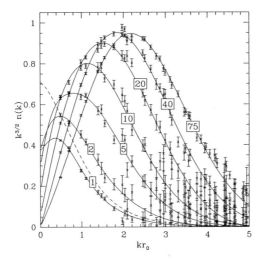

FIGURE 5.8: Momentum distribution $k^{3/2} \cdot n(k)$ for bosons at $r_s =$1, 2, 5, 10, 20, 40 and 75 compared with the fits from equation 5.8. The dashed curve gives the perturbative result (5.9) at $r_s = 1$.

large–r tail of $n(r)$, if the improved estimator (5.3) is used. In fact, the quantity

$$m(r) = \frac{n(r)}{1 + (\Gamma(1/4)/4\Gamma(3/4))\sqrt{r_s/r}} \tag{5.6}$$

reaches its asymptotic value $m(r = \infty) = n_0$ at quite small values of r (see Figure 5.5).

In the opposite, large–k limit the cusp condition [120] gives

$$n(k \to \infty) \to \frac{4r_s^2 g(0)}{(kr_0)^6} \tag{5.7}$$

where $g(0)$ is the static pair correlation function at zero separation, which was independently sampled by DMC.

We now present our simulation results and the analytical fit, which incorporates the above asymptotic behaviours. Our simulation results for the one–body density matrix $n(r)$ are given in Figure 5.7, while Figure 5.8 contains the results for the momentum distribution $n(k)$ and the fit discussed below.

At small density we expect the momentum distribution to be approximately gaussian, in agreement with harmonic theory for the crystalline phase. We have therefore chosen the following expression to interpolate the DMC data:

$$n(k) \;=\; (2\pi)^2 n n_0 \delta^2(k) + \frac{n_0\sqrt{r_s/2}}{k^{3/2}}e^{-k^2/2x_4^2}$$

TABLE 5.4: Best fit parameters for equation (5.8). The last line reports the value of $g(0)$ from Figure 5.2 as used in the fit of $n(k)$.

r_s	2	5	10	20	40	75
n_0	0.373	0.183	0.0748	0.018	0.000528	0.00111
x_0	0.705	1.35	1.33	1.02	0.685	0.403
x_1	0.115	0.166	-0.0181	-0.203	-0.106	-0.0355
x_2	-1.76	-1.58	-0.697	0.379	0.997	1.38
x_3	1.49	1.53	1.51	1.74	1.6	1.5
x_4	-0.609	-0.421	-0.471	-0.609	-0.564	9.52
x_5	8.09	205	–	–	–	–
$g(0)$	0.078	0.01	–	–	–	–

$$+ \left(\frac{x_0}{\sqrt{k}} + x_6 + x_1\sqrt{k} + x_7 k \right) e^{-(k^2 - 2kx_2)/2(x_3^2 + x_4^2)} + \frac{4g(0)r_s^2}{x_5 + k^6} .\quad(5.8)$$

(k and n are in units of r_0^{-1} and r_0^{-2} respectively). Given the known values of the density and of $g(0)$ (see Section 5.1.2), we determined the remaining parameters by a least–squares fit to the DMC data on $n(k)$, $n(r)$ and on the average kinetic energy, as obtained in Section 5.1.1. In order to give the same weight to the real–space, the reciprocal–space and the kinetic energy data we minimized the sum of the reduced χ^2 values, and attributed a 1% error to the kinetic energy. Table 5.4 contains the best–fit parameters and the resulting value of the condensate fraction n_0. The latter is not significantly different from zero at $r_s = 40$ and 75.

Figure 5.8 displays the DMC data, the fit and the curve obtained by perturbation theory (PT) along the lines of Foldy [90], i.e. within the Bogolubov formalism. The result

$$n^{\text{PT}}(k) = \frac{k^2/2m + nv_k - \omega_k}{2\omega_k} \qquad (5.9)$$

with $\omega_k = \sqrt{2nv_k k^2/2m + (k^2/2m)^2}$ and $v_k = 2\pi e^2/k$ is valid in the uniform (or RPA) limit $r_s \to 0$ and compares favourably with the DMC data at $r_s = 1$. With growing r_s the shape of the curves shown in Figure 5.8 changes qualitatively, with exchange effects playing only a minor role at low density (see Figure 5.13 and corresponding discussion in Section 5.3).

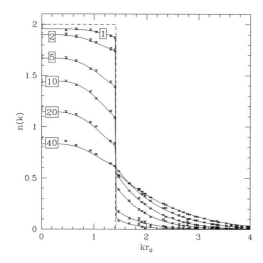

FIGURE 5.9: Momentum distribution $n(k)$ for 2D fermions from DMC (crosses with error bars), interpolation (5.10) (solid line), for r_s =1, 2, 5, 10, 20 and 40 (from upper to lower curve for $k < k_F$, in the opposite order for $k > k_F$).

5.2 2D fermions

5.2.1 Ground state energy

Our results on the ground state energy and the pair correlation function are in very good agreement (within statistical error bars) with those of Rapisarda and Senatore [114], and constitute a check of the correctness of the details of the numerical procedure, which are different in the two computations. The energies for $r_s = 1$ and 2 were not computed by Rapisarda and Senatore. We find $E_g(r_s = 1) = -0.4191(6)$ and $E_g(r_s = 2) = -0.5152(8)$, which are in good agreement with the result obtained from their interpolation $E_g(1) = -0.41953$ and $E_g(2) = -0.5145$ (all energies are in Rydberg per particle). This proves the effectiveness of their interpolation formula in the region between the MC points and the asymptotic $r_s \to 0$ region.

5.2.2 Momentum distribution

Figure 5.9 contains our results for the momentum distribution $n(k)$, and figure 5.10 our results for the one–body density matrix $n(r)$. The crossover from an almost ideal gas situation at $r_s = 1$ to a strongly correlated fluid at $r_s = 40$ is clearly shown. Note that our results for $r_s = 40$ correspond to the metastable unpolarized liquid phase. We collected our results on $n(k)$ and $n(r)$ in a fitting

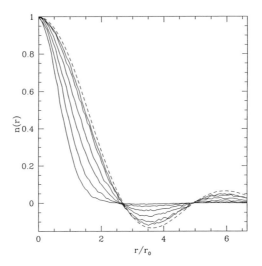

FIGURE 5.10: One–body density matrix $n(r)$ for 2D fermions from DMC for $r_s =1$, 2, 5, 10, 20 and 40 (from upper to lower curve at small r). The dashed line is the noninteracting fermion result $n_0(r) = 2J_1(rk_F)/rk_F$.

TABLE 5.5: Fit parameters and momentum distribution discontinuity Z_F for 2D fermions at $r_s =1$, 2, 5, 10, 20 and 40, from DMC (from equation (5.10)).

r_s	1	2	5	10	20	40
Z_F	1.754	1.553	0.9657	0.5349	0.2059	0.0533
x_0	1.958	1.9	1.67	1.435	1.143	0.8292
x_1	-0.01297	0.00528	0.009721	0.02612	0.007975	0.008649
x_2	0.02773	0.004692	-0.003434	0.108	0.01505	0.00832
x_3	-0.01206	-0.06888	-0.05649	-0.3253	-0.09478	-0.06632
x_4	-0.03796	-0.06144	-0.1666	-0.008001	-0.1839	-0.1415
x_5	8.714	1.811	1.059	1.169	1.163	1.346
x_6	1.745	1.888	1.914	3.66	2.01	2.119
x_7	0.1115	0.4622	0.9791	1.039	0.9777	0.7707
x_8	-0.07	-0.2755	-0.5134	-0.404	-0.3819	-0.26
x_9	0.01101	0.04319	0.08415	0.04138	0.08323	0.07823
x_{10}	0.01039	0.04699	0.09488	0.06176	0.1111	0.08985

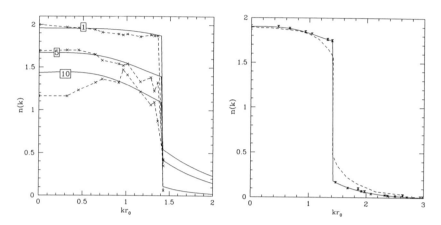

FIGURE 5.11: Left panel: Fit to the DMC momentum distribution $n(k)$ for 2D fermions from equation (5.10) (full curves) compared with the VMC results of Tanatar and Ceperley [115] (crosses joined with dashed lines) at r_s =1, 5 and 10. Right panel: Fit (full curve) and DMC data (crosses with error bars) at $r_s = 2$ compared with the approximate theory of Santoro and Giuliani [123] (dashed curve).

formula incorporating the known large-k behaviour given by equation (5.7)*

$$n(k) = \begin{cases} x_0 + x_1 k + x_2 k^2 + x_3 k^3 + x_4 k^4 + x_{10} k^5 & (k < k_F) \\ \dfrac{4g(0)r_s^2}{k^6} + \left(x_7 + x_8 k + x_9 k^2\right) \exp\left\{ -\dfrac{k - k_F}{x_5} - \dfrac{(k - k_F)^2}{x_6^2} \right\} & (k > k_F) \end{cases}$$

$$(5.10)$$

The kinetic energy, as obtained from the equation of state of RS, was used as an additional data point with an estimated error of 1%. The resulting fit parameters are reported in Table 5.5. The table also contains the resulting discontinuities at the Fermi wavevector Z_F, also called pseudoparticle renormalization factor. While the fit is not completely satisfactory in real space, the good agreement with the k-space data is evident from Figure 5.9.

Figure 5.11 compares our results on $n(k)$ with the VMC results of Tanatar and Ceperley. The drop they observed in $n(k)$ at small k is completely absent in our results. The second part of the same Figure compares our results with the theoretical computation of Santoro and Giuliani at $r_s = 2$ [123], giving good agreement at small k but indicating an overestimate of the reduction of the Fermi surface discontinuity in the Santoro–Giuliani computation.

---

*k is in units of r_0^{-1}, and $k_F r_0 = \sqrt{2}$.

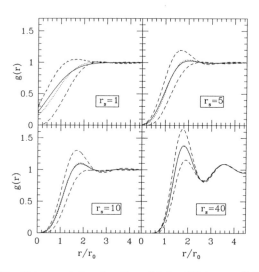

FIGURE 5.12: Pair correlation function $g(r)$ for 2D bosons (full curves) and fermions at different r_s values. The dotted curve gives the spin–averaged $g(r)$, the dashed curves the spin–resolved ones, parallel spins corresponding to a smaller $g(r)$ at intermediate r.

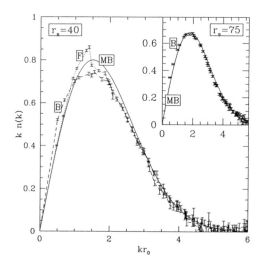

FIGURE 5.13: Momentum distribution $n(k)$ for 2D fermions (F) and bosons (B) from DMC (crosses with error bars), interpolations (5.10) and (5.8) (solid lines) and for classical particles with a fictitious temperature (dashed curve), at $r_s = 40$ and $r_s = 75$ (inset, bosons only).

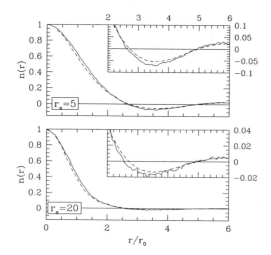

FIGURE 5.14: One particle density matrix for fermions from DMC (solid curves) at $r_s = 5$ and 20 compared with the factorization $n_B(r)n_F^0(r)$ (dashed curves).

5.3   Comparison between bosons and fermions

At low density statistics plays a minor role. Figure 5.12 compares the bosonic and fermionic pair correlation functions $g(r)$ at different r_s. While the spin–resolved fermionic $g(r)$ are rather different due to local antiferromagnetic ordering, the spin–averaged one is very close to the bosonic one for $r_s \geq 5$.

Figure 5.13 compares the momentum distribution $n(k)$ at $r_s = 40$ and 75 with the Maxwell–Boltzmann distribution corresponding to classical particles with a fictitious temperature T_{eff}. This result is consistent with the small differences in energy between the different fluid phases shown in Figure 5.1. The significant difference in crystallization density between bosons and fermions results from a competition between the small extra kinetic energy cost of the Fermi liquid with respect to the Bose liquid and the small potential energy gain of the solid with respect to a strongly correlated liquid.

It was recently proposed [124] that the one–body density matrix for fermionic systems can be approximately factorized as

$$n_F(r) = n_F^0(r)n_B(r) \tag{5.11}$$

where $n_B(r)$ is the corresponding bosonic density matrix and $n_F^0(r) = 2J_1(rk_F)/rk_F$ is the one of noninteracting fermions. This factorization was verified to be approximately satisfied by MC estimates for $n(r)$ in Helium-3 in Ref. [124]. Figure 5.14 compares the DMC data at $r_s = 5$ and 20 with equation (5.11), showing that this approximate factorization is well satisfied also in charged systems.

5.4   Conclusions

In summary, we presented DMC results on the ground state energy, structure factor and momentum distribution of 2D bosons and fermions. Our results on the ground state energy of bosons and on the momentum distribution for both statistics have been interpolated with simple analytical formulas incorporating known asymptotic behaviours. The qualitative picture is similar to the one discussed for the 3D case in the previous chapter: quantum statistics is dominant at small r_s, but at small density its effects are overcome by correlation, which prevents particles from being at contact. In fact, the pair correlation function at zero separation $g(0)$ is almost undistinguishable from zero at $r_s = 10$ or higher. At the same time the Fermi surface discontinuity Z_F for fermions and the condensate fraction n_0 for bosons are strongly reduced with respect to the noninteracting value assumed for $r_s \to 0$.

Part III

Correlations in 2D electron systems

Chapter 6

Charge transfer instability in the bilayered electron gas

6.1 Introduction

Recent experimental developments brought attention to multilayered bidimensional systems, where correlations are enhanced with respect to single layers and additional degrees of freedom are obtained tuning the layer separation d and varying separately the charge present in each layer. The electron–hole bilayer was predicted long ago to have a superfluid low–temperature phase [125], but no conclusive experimental evidence has been obtained so far. A number of different phase transitions has been observed in electron bilayers in the high magnetic field Fractional Quantum Hall regime [126, and references therein]. Enhanced exchange and correlation effects in bilayered 2D electron gas (2DEG) systems have been studied by various authors, leading to the prediction of a lower critical density for Wigner crystallization and to the proposal of some novel interaction–driven phases [114,127–133].

Based on the well-known fact that the compressibility of the low density 2DEG is negative, Ruden and Wu (RW) [127] argued that exchange and correlation could overcome the kinetic and Hartree energy costs stabilizing a charge transfer state (CTS), in which one layer contains all the electrons and the other is empty. Their Hartree–Fock (HF) computation, restricted to the ideal zero-tunneling symmetric case, predicts stability of the CTS for

$$r_s > \frac{3\pi(\sqrt{2}+1)}{16}\left(1+2\frac{d}{a_B}\right) \simeq 1.42\left(1+2\frac{d}{a_B}\right), \tag{6.1}$$

which is well into the range of the experimentally attainable electron densities for reasonable layer separations d (e.g., with GaAs parameters, $r_s = 4$ or $n_{2D} = 2 \cdot 10^{10} \text{cm}^{-2}$ for $d < 90$Å, according to Ref. [127]). In fact higher values of the coupling ($r_s \approx 20$) have been achieved [134] working with holes, rather than electrons. Note that in the bilayer we relate the density/coupling

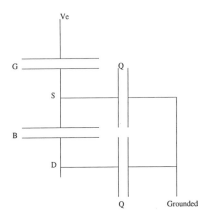

FIGURE 6.1: Equivalent circuit model from Ref. [139]. The external potential is applied at the top gate, with the two 2DEG's grounded. The capacitor G represents the barrier between the gate and the first 2DEG, the capacitor B represents the barrier between the two quantum wells, and the two capacitors Q represent the so-called "quantum capacitance" $C_Q = m^* e^2 / \pi \hbar^2$ of the two 2DEG's.

parameter r_s to the total areal density, $n_{2D} = 1/\pi r_s^2 a_B^2$, with a_B the Bohr radius.

Evidence for a CTS in GaAs/AlAs double quantum well structures in the presence of bias voltage was recently reported [135–137]. The experimental setup allowed to measure independently the charge contained in the two layers as a function of bias voltage, via Shubnikov–de Haas oscillations. For low voltages the transferred charge was found to depend linearly on V^e, in quantitative agreement with the predictions of a simple equivalent circuit model (see Figure 6.1), but for high enough voltages $V_g \simeq 1V$ the charge of one layer dropped abruptly from around 20% of the total charge to 0. In the next section we prove that, for two dimensional electron layers at zero bias voltage, the CTS is unstable at any density and layer separation d [138]. The interpretation of the experimental data will be discussed in Section 6.3.

6.2 Zero–temperature phase diagram

A convenient model Hamiltonian for the coupled electron layers is obtained introducing a pseudospin variable τ_i, which labels the planes with the convention that $\tau_i^{(z)} = 1/2\ (-1/2)$ if electron i is in the upper (lower) plane. For fixed interlayer distance and background densities and in the absence of tunneling, the Hamiltonian takes the form

$$H = \sum_i \frac{p_i^2}{2m} + \frac{1}{2}\sum_{i\neq j}' \left(\frac{e^2}{r_{ij}} \frac{1 + 4\tau_i^{(z)}\tau_j^{(z)}}{2} + \frac{e^2}{\sqrt{d^2 + r_{ij}^2}} \frac{1 - 4\tau_i^{(z)}\tau_j^{(z)}}{2} \right) + V^e \sum_i \tau_i^{(z)},$$

(6.2)

where V^e is the energy difference between the two planes, given by $2\pi e^2(\rho_1 - \rho_2)d$ in the absence of external fields, ρ_1 and ρ_2 being the background densities of the two planes. The prime on the second sum prime indicates that H is regularized in the thermodynamic limit with a τ – independent term representing interactions with a symmetric uniform background, $\rho_s = (\rho_1 + \rho_2)/2$ in each layer. With N electrons there are two CTS's, corresponding to the two eigenstates of the total pseudospin component along z, $\mathcal{T}^{(z)} = \sum_i \tau_i^{(z)} = \pm N/2$. In other words CTS is the same as (pseudospin) ferromagnetism, with the magnetization oriented along the z axis. We shall show that without bias voltage there can be no such ferromagnetic phase transition. We will further argue that in the presence of tunneling a transition takes place to a (coherent) correlated state, which has the same symmetry as the CTS but does not displace any charge.

It is interesting to consider the ideal case $d = V^e = 0$, where both real spin σ and pseudospin τ are conserved and the system is actually a 4–component monolayer 2DEG. At large density (small r_s) the ground state is completely paramagnetic, with the four components equally populated to minimize the Fermi energy; in the opposite low density regime the ground state is known to be a Wigner crystal and essentially insensitive to spin polarization. In the intermediate regime the 4–component 2DEG can be expected to mimic the above–mentioned behavior of the 2DEG (and 3DEG), with the appearance of a spin–polarized ground state. In HF, the 4–component 2DEG is unstable towards the 2–component phase at $r_s = 3\pi(\sqrt{2} + 1)/16 \simeq 1.42$ and then towards the 1–component phase at $r_s = 3\pi(2 + \sqrt{2})/16 \simeq 2.01$, the first being obviously identical with the $d \to 0$ limit of the RW results and the latter with the usual spin polarization transition of the 2DEG in HF.

To assess the actual presence of these phase transitions we performed Slater–Jastrow Variational and Fixed Node Diffusion Monte Carlo simulations for the 4–component 2DEG, and compared them with the results of Rapisarda and Senatore (RS) [114] on the unpolarized and spin–polarized 2DEG. Our results are reported in Table 6.1, and compared in Fig. 6.2 with the data by RS on the 2 and 1–component 2DEG. At variance with the simple HF prediction, the figure clearly shows that the 4–component phase crystallizes at $r_s \simeq 42$, being always stable at smaller couplings with respect to the other two fluid phases that we have considered. We do not expect backflow corrections to spoil the validity of our results since they are known to lower the paramagnetic energies more than the ferromagnetic ones.

As we have already mentioned, a total charge transfer state is described

TABLE 6.1: Fixed–node DMC total energies of the 4–component 2DEG for 52 particles and in the bulk limit, in Ryberg per particle. Variational results used for the size–extrapolation are also given. These data are accurately reproduced by the same fitting formula used by RS, with parameters $a_0 = -0.88115$, $a_1 = 4.439$, $a_2 = 0.14063$, $a_3 = 1.9034$.

	$r_s = 2$	$r_s = 10$	$r_s = 20$	$r_s = 30$	$r_s = 50$
$N = 36$	-0.5882(4)	-0.17102(2)	-0.092275(6)	-0.063561(5)	-0.039386(3)
$N = 52$	-0.5728(1)	-0.17030(2)	-0.092085(4)	-0.063474(3)	-0.039354(1)
$N = 84$	-0.58062(7)	-0.17055(2)	-0.092123(5)	-0.063481(4)	-0.039355(2)
$N = 100$	-0.5758(1)	-0.17036(2)	-0.092062(4)	-0.063457(4)	-0.039344(2)
$N = 180$	-0.57676(8)	-0.17031(2)	-0.092050(5)	-0.063453(4)	-0.039340(2)
VMC_∞	-0.5750(1)	-0.17019(3)	-0.092002(4)	-0.063426(3)	-0.039330(1)
DMC_{52}	-0.5838(4)	-0.17232(4)	-0.092891(9)	-0.063975(8)	-0.039639(3)
DMC_∞	-0.5860(5)	-0.17221(6)	-0.09281(1)	-0.063927(9)	-0.039615(4)

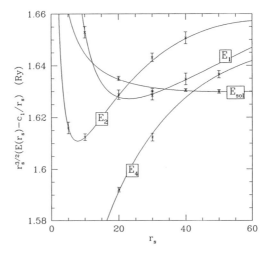

FIGURE 6.2: Ground state energy per particle as a function of r_s for the normal unpolarized (E_2), polarized (E_1) and 4–component (E_4) 2DEG and for the triangular Wigner crystal (E_{sol}). Except for E_4, all data are from RS. On the purpose of clarity we plotted $r_s^{3/2}(E(r_s) - c_1/r_s)$, with $c_1 = -2.2122$.

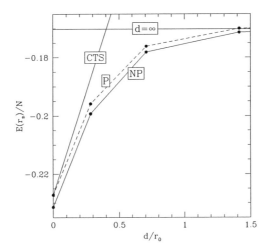

FIGURE 6.3: Ground state energy per particle of different phases as a function of d/r_0 at $r_s = 10/\sqrt{2} \simeq 7.1$. The $d = \infty$ curve gives the value $E_2(r_s\sqrt{2})$ for uncorrelated planes and the CTS curve gives the value $E_2(r_s) + (\pi/2)e^2 d\rho_{2D}$ for the total (unpolarized) charge transfer state. NP (P) indicates the ground state energies of Rapisarda and Senatore [114,132] for the normal unpolarized (spin–polarized) phase at various distances d (error bars are smaller than symbols). The points from RS are joined with $E_4(r_s)$ ($E_2(r_s)$) at $d = 0$ which is the corresponding zero–separation limit if tunneling is neglected. Stability of the NP phase is evident.

in our formalism as an eigenstate of each $\tau_i^{(z)}$ with eigenvalue $+1/2$. We remark that $\tau_i^{(z)}$ commutes with the Hamiltonian even at finite d and V^e (but zero tunneling) and we label with $|\Psi_{\text{ch.tr.}}\rangle$ the lowest eigenstate within the charge transfer subspace. In the symmetric $V^e = 0$ case it can be easily seen that the coherent state $|\Psi_{\text{coh.}}\rangle = \exp\left\{i\frac{\pi}{2}\sum_j \tau_j^{(y)}\right\}|\Psi_{\text{ch.tr.}}\rangle$ has an average energy lower than $|\Psi_{\text{ch.tr.}}\rangle$. In fact, (i) $1/x > 1/\sqrt{d^2 + x^2}$ for any x; and (ii) $\langle\Psi_{\text{ch.tr.}}|4\tau_i^{(z)}\tau_j^{(z)}|\Psi_{\text{ch.tr.}}\rangle = 1$, whereas $\langle\Psi_{\text{coh.}}|4\tau_i^{(z)}\tau_j^{(z)}|\Psi_{\text{coh.}}\rangle = 0$. The variational principle implies that $|\Psi_{\text{ch.tr.}}\rangle$ cannot be the ground state of the system, even if $|\Psi_{\text{coh.}}\rangle$ is not an eigenstate of H.

We stress that both $|\Psi_{\text{ch.tr.}}\rangle$ and $|\Psi_{\text{coh.}}\rangle$ have the same symmetry properties under exchange of the particles' spatial coordinates, as this is not affected by pseudospin rotation. In particular the two states are both fully pseudospin polarized, though in different directions. The average charge distribution in $|\Psi_{\text{coh.}}\rangle$ corresponds to half electrons in one layer and half in the other, and therefore it does not loose in Hartree energy with respect to the normal state, still having the exchange energy gain deriving from the full antisymmetry of the spatial part of the wavefunction. The difference with the above–mentioned HF predictions is due to the fact that no state with the same symmetry as $|\Psi_{\text{coh.}}\rangle$ was considered by RW. Additional evidence for instability of the CTS is given, on the basis of DMC simulations, in Figure 6.3. With reference to the figure, the RW computation is equivalent to a comparison of the HF approximation to the $d = \infty$ and CTS curves alone.

No such clearcut statement can be made in the asymmetric $V^e \neq 0$ case. The term $V^e \sum_i \tau_i^{(z)}$ clearly favours finite values of $\mathcal{T}^{(z)}$, $i.e.$ (partial) pseudospin polarization, and can possibly stabilize the fully polarized phase, as we discuss below in the large d case.

It must be noticed that the state $|\Psi_{\text{coh.}}\rangle$ is a broken symmetry state, since all states of the form

$$|\Psi_{\text{coh.}}(\phi)\rangle = \exp\left\{i\frac{\pi}{2}\sum_j(\tau_j^{(y)}\cos\phi + \tau_j^{(x)}\sin\phi)\right\}|\Psi_{\text{ch.tr.}}\rangle \qquad (6.3)$$

are degenerate. The order parameter is the total pseudospin $\mathcal{T} = \sum_i \tau_i$, corresponding to interlayer phase coherence, and behaves as an easy–plane ferromagnet, the z component being frozen by the interplay of V^e and the Hartree energy. Most of the considerations presented in Ref. [126] for the high B situation can be directly extended to this zero field case. The role of a tunneling Hamiltonian, which can be conveniently written as $H_t = t\sum_i \tau_i^{(x)}$, with the tunneling matrix element $t > 0$, is to break the ϕ symmetry and stabilize the coherent state $|\Psi_{\text{coh.}}\rangle = |\Psi_{\text{coh.}}(\phi = 0)\rangle$, in which $\langle\mathcal{T}^{(z)}\rangle = 0$ and $\mathcal{T}^{(x)} = -N/2$. In the large t case all electrons trivially lie in the symmetric state, which has exactly the same form as $|\Psi_{\text{coh.}}\rangle$. At sufficiently small d, a perturbative estimate of the critical amplitude t^* at which the fully coherent state $|\Psi_{\text{coh.}}\rangle$ becomes stable is easily obtained from the energies of Table 6.1 and of RS for $d = t = 0$. To first order a finite t splits the pseudospin polarized multiplet at energy E_2, lowering $|\Psi_{\text{coh.}}(\phi = 0)\rangle$ to energy $E_2(r_s) - t/2$. The singlet state E_4 is unaffected. Thus one gets $t^*/2 = E_2(r_s) - E_4(r_s)$, which at $r_s = 2$ corresponds to $t^* = 0.8\,\text{meV} = 9\,\text{K}$ with GaAs parameters.

6.3   Connection with experimental data

The experimental setup of Ref. [135–137] includes a metallic gate at a large distance D from the two 2DEG's and some charged dopants, whose location and amount are unspecified and irrelevant for the present purposes. The 2DEG's are realized with 150(180)Å thick GaAs quantum wells with a midpoint separation d of 220(194)Å for sample A(B), and the tunneling energy $t = 0.005K\,(5K)$. From the work of Zheng and MacDonald [140] and from numerical simulations [132] it is clear that interlayer correlations are relevant only if d becomes smaller than the average interparticle distance $r_s a_B$. As in the present situation $r_s a_B \sim 150$Å and $d \sim 200$Å correlation effects can be

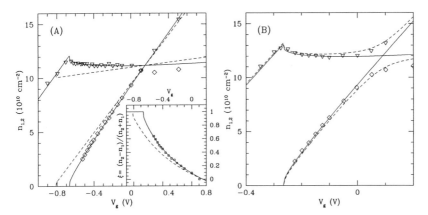

FIGURE 6.4: Layer densities for the bilayers A and B specified in the text, as a function of bias voltage V_g. Triangles and diamonds give the experimental results of Ref. [136], and the full curve results from the strictly two dimensional model discussed in the text. In (A), the dashed curve is obtained neglecting both correlation and exchange, while the inset reports the pseudospin polarization ξ as a function of V_g. In (B), the dashed curve is obtained by accounting for tunneling as explained in the text.

safely neglected. In the following we shall further neglect finite thickness and tunneling effects.

We propose a simple model which correctly includes intralayer correlations and interlayer mean field interactions and is closely related to the one used by Eisenstein *et al* [141] to discuss a similar experimental setup. Let us consider the gate potential $V_g^{(0)}$ corresponding to equal density $n^{(0)}$ in the two 'active' layers, *i.e.* to zero electric field between the layers. [The value of $V_g^{(0)}$ depends on details of the sample design, as location and amount of dopants, geometry, etc.]. With reference to this state and under the assumption that no net charge flows in the system with varying V_g, the electrochemical equilibrium condition between the two 2DEG's is written as

$$\mu(n_1) - \mu(n^{(0)}) = \mu(n_2) - \mu(n^{(0)}) + 4\pi e^2 d(n_2 - n_0) \qquad (6.4)$$

and between the upper layer and the gate as

$$-e(V_g - V_g^{(0)}) = \mu(n_1) - \mu(n^{(0)}) + 4\pi e^2 D(n_1 + n_2 - 2n^{(0)}) \qquad (6.5)$$

We now proceed to solve numerically these two coupled equations for the unknowns n_1 and n_2 at various values of V_g. While the sample parameters d, $n^{(0)}$ and $V_g^{(0)}$ are given in Ref. [136], D is not, and was determined by fitting the measured variation of the average density $(n_1 + n_2)/2$ with V_g. The chemical potential $\mu(n) = d(nE(n))/dn$ for the idealized single layer 2DEG is derived from the precise equation of state obtained by RS with DMC. The results shown in Fig. 6.4 are in very good agreement with the experimental data from Ref. [136], with significant disagreement only in the tunneling–dominated region around $n_1 = n_2$. In fact quantitative agreement (see Fig. 6.4.B) can be obtained even in this region computing the subband densities from the layer densities with no tunneling $\tilde{n}_1$, $\tilde{n}_2$ by exact diagonalization of a noninteracting electron hamiltonian with finite tunneling,

$$H = \sum_k \left[\left(\frac{\hbar^2 k^2}{2m} - \mu^0(\tilde{n}_1) \right) a_{k,1}^\dagger a_{k,1} + \left(\frac{\hbar^2 k^2}{2m} - \mu^0(\tilde{n}_2) \right) a_{k,2}^\dagger a_{k,2} + t \left(a_{k,1}^\dagger a_{k,2} + h.c. \right) \right].$$
(6.6)

We remark that neglecting intralayer interactions would lead to the usual approximation $\mu^0(n) = C_q n$, where the so–called quantum capacitance C_q is given by $\pi \hbar^2 / m$. By making this assumption for both layers the above equations are linear and can be solved analytically, obtaining results in fair agreement with experiments but completely missing the nontrivial charge transfer effect signaled by the peak in the upper curve of Fig. 6.4.

6.4   Conclusions

In conclusion, we presented a pseudospin formalism to describe bilayer 2DEG systems and used it to rigorously prove the instability of the charge transfer state and to argue for the stabilization of coherent states at small finite tunneling. We also reported results from DMC simulations, which reveal the stability of the pseudospin unpolarized state at $d = V^e = 0$. Finally we have shown that the presently available experimental data, being in a regime where interlayer correlations are negligible, can be quantitatively accounted for by a simple strictly two dimensional model. Clearly the more challenging and interesting situations, in which interlayer correlations are discernible and determine the details of the charge transfer, is still awaiting for both experimental and theoretical investigations.

Chapter 7

Collective Modes and Electronic Spectral Function in Smooth Edges of Quantum Hall Systems

7.1 Introduction

Since the discovery of the Quantum Hall (QH) Effect 2D electrons in high magnetic fields have been one of the most studied examples of strongly interacting many–body systems [142–144], and still pose a number of intriguing questions. Understanding the character of edge excitations is crucial to the theory of transport properties of quantum Hall bars, quantum wires and dots [145–148].

The edge of a confined QH fluid is a one dimensional system, motion in the transverse direction being suppressed by bulk incompressibility. Furthermore, the two propagation directions along the edge are inequivalent, due to the presence of the magnetic field. An effective theory of edge excitations was first derived by Wen [149–151], who showed that a "sharp" edge is a realization of the one-dimensional chiral Luttinger liquid (CLL) model. Wen's theory predicts that the electronic spectral function exhibits a nontrivial behavior, leading to a density of states that vanishes, at low energy, as a power law. This theory has been well confirmed by detailed microscopic calculations [152] and by recent experiments at $\nu = 1/3$ [153,154]. The situation at other filling factors is more complex, and in particular for $\nu = 1/2$ there is a significant disagreement between existing theories based on the Composite Fermion picture [155], and experiment [156]. The effect of the long range of the Coulomb interaction, which was initially ignored, has been recently included by Zülicke and MacDonald [157].

All the above papers assumed the validity of the so-called "sharp edge" model, in which the density of the system drops sharply from the bulk value ρ_0 to zero within a few magnetic lengths $\ell = (\hbar c / eB)^{1/2}$. There are numer-

ous indications that this is not always the case. On one hand, Hartree-Fock
calculations [158] for strongly confined systems predict that, at sufficiently
strong magnetic field, the edge undergoes a reconstruction and has a more
extended shape. On the other hand, in the case of *smooth* confinement, such
as can be realized by gate electrodes, the electronic density is expected to have
a smoothly varying profile (on the scale of ℓ), determined by classical elec-
trostatic equilibrium [159–161]. Detailed calculations using density functional
theory, Thomas-Fermi theory, and other methods [162–166] have confirmed the
theoretical validity of this "smooth edge" picture. Edge imaging experiments
[167] have confirmed the relevancy of this description for gate-confined Hall
bars.

   This Chapter presents a study of the spectral properties of the "smooth
edge" model, where the mapping to a one dimensional chiral electron liquid is
not justified. In fact, a recent study by Aleiner and Glazman (AG) [168,169]
based on the classical hydrodynamics approach has shown that a smooth edge,
in contrast to a sharp edge, supports *multiple branches* of edge waves. Of these,
one is the usual edge magnetoplasmon [170,171], and the others (infinitely
many in the classical approach of AG) are phonon-like and lower in energy than
the magnetoplasmon. We shall show that, under the assumption of smooth
density variation, only a finite number of these phonon-like modes are correctly
described as independent bosons.

   In order to calculate the electronic spectral function we rely on the strong
analogy between this problem and that of a uniform electron gas in the par-
tially filled lowest Landau level (LLL). In both cases the self–consistent mean–
field potential is uniform, so that the electrons are distributed among a large
number of degenerate orbitals at the Fermi energy. In a smooth edge this
occurs because the nonuniform electronic density perfectly screens the field
due to the external confinement potential [148,161]. Recently, Johansson and
Kinaret [172] have shown that a qualitatively correct description of the spectral
function [173] of the uniform electron gas in the LLL at general filling factor is
given by an independent boson model (IBM) [174] which describes the interaction of a single localized electron with the density fluctuations of the system.
An essentially equivalent procedure has been applied by Aleiner, Baranger and
Glazman [175,176] to study the spectral function of the 2D electron liquid in a
weak magnetic field. Finally, a formal justification of the IBM from diagrammatic many-body theory has been provided by Haussmann [177]. Encouraged
by these successes, here we apply the independent boson model to the problem
of the smooth edge. In the limiting case of a sharp edge the resulting theory
reduces to standard bosonization in the cases where there is only one branch
of edge waves, and is in good agreement with experiment [178,154,156], which
has been performed only in this regime. When multiple branches are present,
our results for the low energy behavior of the tunneling density of states are
significantly different from those of the sharp edge model [179]. The actual

number of modes that must be included depends on the width of the edge, as explained below.

7.2   Collective edge excitations

Let us begin by writing down the microscopic Hamiltonian within the LLL in terms of density fluctuations relative to the equilibrium density profile $\rho_0(y)$:

$$H = \frac{1}{2} \int_{\text{edge}} \frac{e^2}{|\mathbf{r} - \mathbf{r}'|} \delta\rho(\mathbf{r})\delta\rho(\mathbf{r}')d^2\mathbf{r}d^2\mathbf{r}', \qquad (7.1)$$

where the density operator (projected in the LLL) has been written as $\rho(\mathbf{r}) = \rho_0(y)+\delta\rho(\mathbf{r})$. The integral in equation (7.1) extends over the edge region which we take to be $0 < x < L$, $0 < y < d$, with $L \gg d \gg \ell$, and translationally invariant along x ($\rho_0(y) = 0$ for $y < 0$). The projected density fluctuation operator is given by

$$\delta\rho(\mathbf{r}) = \frac{1}{\sqrt{\pi}\ell L} \sum_{h \neq k} c_k^\dagger c_h e^{i(h-k)x} e^{-\frac{(y-y_k)^2+(y-y_h)^2}{2\ell^2}}, \qquad (7.2)$$

where $y_h = h\ell^2$, h and k are integral multiples of $2\pi/L$, $c_k^\dagger$ is the creation operator of a Landau gauge orbital centered a y_k with wave vector k in the x direction. Note the restriction $h \neq k$ which excludes the equilibrium component of the density. The kinetic energy is absent in equation (7.1) due to projection on the LLL, and we have assumed that density fluctuations vanish in the bulk of the system, i.e., the bulk is incompressible. The terms linear in $\delta\rho$ have vanished because of the equilibrium condition $\int \rho_0(\mathbf{r})v(\mathbf{r} - \mathbf{r}')d^2\mathbf{r}' + V_{\text{ext}}(\mathbf{r}) = constant$ where $V_{\text{ext}}(\mathbf{r})$ is the confinement potential. Therefore the problem is formally similar to that of a translationally invariant electron gas: the nonuniformity enters only through the restricted region of integration in equation (7.1).

The normal mode operators $\delta\rho_{nk}$ are now introduced according to the definition

$$\delta\rho_{nk} = \int_0^L \frac{dx}{L} e^{-ikx} \int_0^d dy f_{nk}(y)\delta\rho(x,y), \qquad (7.3)$$

where $f_{nk}(y)$ are the solutions of the equation

$$\int_0^d K_0(k|y - y'|) f_{nk}(y') \frac{\rho_0'(y')}{\bar{\rho}} dy' = \frac{1}{\lambda_{nk}} f_{nk}(y), \qquad (7.4)$$

where $K_0(y)$ is the modified Bessel function. They satisfy the orthonormality condition $\int_0^d f_{nk}(y) f_{mk}(y) \frac{\rho_0'(y)}{\bar{\rho}} dy = \delta_{nm}$, and vanish outside the interval $[0, d]$. Equation (7.4) is the eigenvalue problem solved by AG, the collective excitations frequencies ω_{nk} being given by $k\bar{\nu}e^2/\lambda_{nk}\pi$, where $\bar{\nu} = 2\pi\ell^2\bar{\rho}$

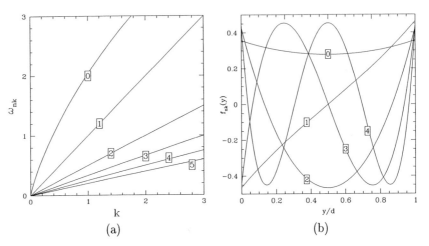

FIGURE 7.1: Collective excitation frequencies ω_{nk} and form factors in the transverse direction $f_{nk}(y)$ for the first few modes corresponding to the edge profile specified in the text.

is the usual filling factor in the bulk. Figure 7.1 displays the mode dispersion ω_{nk} and the form factors $f_{nk}(y)$ for the first few modes, for the case $\rho_0(y) = (2/\pi)\bar{\rho}$ atn$\sqrt{y/d}$ considered by AG for a gate-confined electron gas. In general, the n-th eigenfunction has n nodes in the y direction, and for $n \geq 1$ the long–wavelength dispersion is acoustic. The edge magnetoplasmon mode ($n = 0$) is the only charged mode, and has a $k \log k$ dispersion due to the long range of the Coulomb interaction.

In terms of the normal modes, the hamiltonian (7.1) takes the form

$$H = \sum_{nk>0} \hbar\omega_{nk} b_{nk}^{\dagger} b_{nk}, \qquad (7.5)$$

where the operators b_{nk} are defined via $\delta\rho_{nk} = \sqrt{k\ell^2 \bar{\rho}/L} b_{nk}^{\dagger}$.

It remains to be determined under what conditions the operators b_{nk} are good boson operators. To this end we substitute equation (7.2) in equation (7.3), noting that when $n \ll d/\ell$ the gaussian factors in the integral can be replaced by δ-functions on the scale of variation of f_{nk}. We obtain

$$\delta\rho_{nk} \simeq e^{-k^2\ell^2/4} \frac{1}{L} \sum_{h} c_{h-k/2}^{\dagger} c_{h+k/2} f_{nk}(h\ell^2). \qquad (7.6)$$

($n \ll d/\ell$).

The commutator of two density fluctuations can now be easily calculated to be

$$[\delta\rho_{nk}, \delta\rho_{m-k}] = e^{-k^2\ell^2/2} \frac{1}{L^2} \sum_{h} (n_{h-k/2} - n_{h+k/2}) \times f_{nk}(h\ell^2) f_{mk}(h\ell^2)$$

$$\simeq -\frac{k\ell^2\bar{\rho}}{L}\delta_{nm},\qquad(7.7)$$

in agreement with the commutation rules for bosons. In arriving at equation (7.7) we have assumed $k\ell \ll 1$, and we have replaced the occupation number operators by their ground-state expectation values n_k, which amounts to a linearization of the equations of motion around the equilibrium state. For the commutator $[\delta\rho_{nk}, \delta\rho_{m-k'}]$ with $k \neq k'$, we find, at the same level of approximation, zero. This is because the commutator in question contains terms of the form $c_h^\dagger c_{h'}$ with $h \neq h'$, which vanish upon averaging in a translationally invariant (along x) state*.

7.3   Electronic spectral function

Having thus completed the bosonization of the hamiltonian, we proceed to the calculation of the spectral function within the independent boson model [172,174]. The model describes a single electron, localized at point $\mathbf{r}$, electrostatically coupled to density fluctuations:

$$H_{IBM} = \sum_{n,k>0} \hbar\omega_{nk}b_{nk}^\dagger b_{nk} + \psi^\dagger(\mathbf{r})\psi(\mathbf{r})\sum_{n,k>0} M_{nk}(y)[b_{nk}^\dagger e^{ikx} + b_{nk}e^{-ikx}],\quad(7.8)$$

where the matrix element $M_{nk}(y)$ is given by

$$M_{nk}(y) = \frac{2e^2}{\lambda_{nk}}f_{nk}(y)\sqrt{\frac{k\ell^2\bar{\rho}}{L}},\qquad(7.9)$$

and $\psi^\dagger(\mathbf{r})$ is the field operator that creates an electron in the LLL coherent state (gaussian) orbital centered at $\mathbf{r}$. The hamiltonian (7.8) can be solved by standard methods [174], within the one-electron Hilbert space. The fermionic Green's function is obtained as

$$\begin{aligned}
G_>(y;t) &= -i\langle\psi(\mathbf{r},t)\psi^\dagger(\mathbf{r},0)\rangle\\
&= (1-\nu_0(y))\exp\left(\sum_{n,k>0}\frac{M_{nk}^2(y)}{\omega_{nk}^2}[e^{-i\omega_{nk}t}-1]\right),\quad(7.10)
\end{aligned}$$

where $\nu_0(y) = 2\pi l^2\rho_0(y)$, and the sum over n and k in the exponent is restricted by the conditions $n \ll d/\ell$ and $k \ll 1/\ell$, which define the regime of validity of the hydrodynamic approximation. This result can also be obtained from

---

*Han and Thouless [180] have argued that the hamiltonian (7.5) should include an additional term describing the dynamics of the boundary between the compressible edge and the incompressible bulk. However, since the additional term is quadratic in the boson operators, the modified hamiltonian can be recast (after a unitary transformation) in the form of equation (7.5), with modified, but qualitatively similar, eigenfrequencies.

direct bosonization of the electron field operator, as is usually done in the CLL picture:

$$\Psi^\dagger(\mathbf{r}) = \exp\left\{ - \sum_{n,k>0} \frac{f_{nk}(y)}{\sqrt{k\ell^2 \tilde{n}L}} \left(e^{ikx} b_{nk}^\dagger - e^{-ikx} b_{nk} \right) \right\}. \tag{7.11}$$

The Fourier transform of $G(y,t)/2\pi$ is the spectral function $A_>(y,\omega)$ and gives the local density of states, which controls the tunneling current from a point contact located at position y into the edge. From equation (7.10) it can be easily shown [181] that $A_>(y,\omega)$ satisfies the integral equation

$$\omega A_>(y,\omega) = \int_0^\omega g(\Omega) A_>(y,\omega - \Omega)d\Omega, \tag{7.12}$$

where

$$g(y,\Omega) = \sum_{n,k} \frac{M_{nk}(y)^2}{\omega_{nk}} \delta(\Omega - \omega_{nk}). \tag{7.13}$$

Equation (7.12), together with the conditions $A_>(y,\omega) = 0$ for $\omega < 0$ and $\int_0^\infty A_>(y,\omega)d\omega = 1 - \nu_0(y)$, completely determines the spectral function. This equation further implies that, at sufficiently small ω, $A_>(y,\omega)$ will have a power-law behavior

$$A(y,\omega) \sim \omega^{g(y,0)-1} \tag{7.14}$$

if and only if the function $g(y,\Omega)$ has a finite limit for $\Omega \to 0$. The tunneling current I, as a function of voltage V, will then behave as $V^{g(y,0)}$ for sufficiently low voltage. Notice that this conclusion is completely general, and does not depend on the specific (hydrodynamic) model that led to the definition of $g(y,\Omega)$ in equation (7.13). In the general case, $g(y,\Omega)$ could be computed from the microscopic density-density response function $\chi_{\rho\rho}(\mathbf{r},\mathbf{r}',\Omega)$ of the edge as follows:

$$g(y,\Omega) = \frac{1}{\Omega} \int d^2\mathbf{r}' d^2\mathbf{r}'' v(\mathbf{r}-\mathbf{r}')v(\mathbf{r}-\mathbf{r}'')\mathrm{Im}\,\chi_{\rho\rho}(\mathbf{r}',\mathbf{r}'',\Omega), \tag{7.15}$$

where $v(\mathbf{r}-\mathbf{r}')$ has the Fourier transform $v(k) = (2\pi e^2/k)\exp\left(-(k\ell)^2/4\right)$ [177]. An important advantage of this microscopic formulation is that the finite lifetime of the collective modes (which is assumed to be infinite in the hydrodynamic model) would be taken into account through the width of the peaks in $\mathrm{Im}\,\chi_{\rho\rho}$.

The calculation of the exponent $g(y,0)$ is easily performed within the hydrodynamic model. Neglecting the weak nonlinearity of the $n = 0$ mode we obtain $g(y,0) = \sum_n \beta_n(y)$, where

$$\beta_n(y) = \frac{1}{\bar{\nu}} f_{n0}^2(y). \tag{7.16}$$

Although the cutoff at $n = d/\ell$ introduces an uncertainty in the evaluation of the exponent at any given d, we emphasize that there would be no uncertainty if one used the microscopic formula (7.15) for $g(y, \Omega)$. Our approximate hydrodynamic evaluation of the exponent should be in good qualitative agreement with the results of the more accurate microscopic calculation.

We observe that independently of the shape of the density profile $\beta_0(y) = 1/\bar{\nu}$, with negligible corrections arising from the weak nonlinearity of the dispersion of the $n = 0$ mode. Therefore in the sharp edge limit, when only one branch of edge waves exists, we recover the familiar result of Wen's theory $A_>(\omega) \sim \omega^{1/\bar{\nu}-1}$. For $n > 1$ $\beta_n(y)$ fluctuates around an average value of $1/\bar{\nu}$ in a way dependent on the form of the equilibrium density profile of the edge[†]. We conclude that the exponent in equation (7.14) increases linearly with d and therefore that in the limit $d \to \infty$ (limit of infinitely smooth edge) the tunneling density of states vanishes at low energy faster than any power law, that is, a "hard" gap develops. However, it is easy to see that the power law behavior of equation (7.14) only holds for $\omega \ll \bar{\nu}e^2/d\pi$ - an interval that shrinks to zero for $d \to \infty$.

In Figure 7.2 we present our numerical results for the full electronic spectral function, calculated from equation (7.12) within the hydrodynamic model for different edge widths d. In contrast to the analysis of the low-frequency behavior, this calculation depends on the detailed form of the eigenfunctions $f_{nk}(y)$ and eigenfrequencies ω_{nk}. From a detailed study of the solutions of the eigenvalue equation (7.4) we have found that the f_{nk}'s can be treated as being independent of k and the ω_{nk}'s to be linear functions of k up to a maximum wave vector $k_c = n/d$ for which the wavelength along the edge equals the wavelength perpendicular to the edge. For $k > k_c$ the mode dispersion becomes approximately constant, and the wavefunction f_{nk} becomes localized near the boundaries of the edge region, giving negligible contribution to the spectral function. The results presented in Fig. 1 have been obtained using the double cutoff $n < d/\ell$ and $k < k_c$: the results are found to be largely independent of the details of the cutoff procedure.

Figure 7.2 shows clearly how the low energy pseudogap becomes more and more pronounced with increasing d (and, therefore, increasing number of branches of edge waves). For very large d the spectral function is found to converge to a δ function centered at $\omega_0 = \bar{\nu}e^2/\pi\ell$, which coincides with the simplest estimate of the potential energy cost for the insertion of an electron into a frozen liquid [177].

Our results for $d \to \infty$ are in qualitative agreement with those obtained in refs. [172] and [177] for the spectral function of the homogeneous electron gas,

---

[†]This can be confirmed by explicit calculation in the special case $\rho_0(y) = (2/\pi)\bar{\rho}$ at $n\sqrt{y/d}$ considered by AG for a gate-confined electron gas, leading to the result $\beta_n(y) = \frac{1}{\bar{\nu}}T_{2n}^2\left(\sqrt{d/(y+d)}\right)(2 - \delta_{n0})$, where $T_n(y)$ is the n-th Chebyschev polynomial.

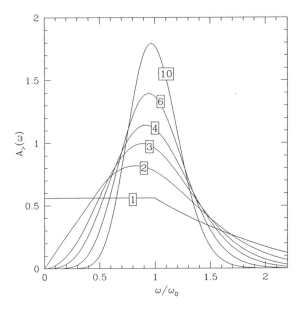

FIGURE 7.2: Electronic spectral function $A_>(\omega)$ as a function of ω/ω_0, where $\omega_0 = \bar{\nu}e^2/\pi l$, for edges of a $\bar{\nu} = 1$ QH system with 1, 2, 3, 4, 6 and 10 modes. Logarithmic corrections to the edge magnetoplasmon dispersion are neglected. The dependence on y and $\rho_0(y)$ have been eliminated neglecting the constant $1 - \nu_0(y)$ and using $\beta_n = 1/\bar{\nu}$, $\omega_{nk} = \bar{\nu}e^2 k/\pi n$. The integral over k has been cut off as explained in the text.

except that the latter is found to have a finite width. This happens because our hydrodynamic approach is unable to give the gapful collective modes of the homogeneous fluid phase [182], and hence our spectral function does not reduce to that of the homogeneous phase.

7.4   Conclusions

In conclusion, we have performed an independent boson model calculation of the tunneling density of states for a smooth edge, and we have found that it vanishes at low frequency as a power, with an exponent that differs significantly from the one found in the sharp edge case. Recent experiments by Chang *et al.* [154] have confirmed the predictions of the CLL for the exponent of the tunneling density of states in a sharp edge at filling factors of the form $1/(2p + 1)$, even if significant disagreement is found at other filling factors [156]. It should be interesting to extend these studies to see if and how the exponents change as the smoothness of the edge is varied.

Summary and concluding remarks

In Part I we studied dynamical correlations in quantum charged fluids at long wavelength, with attention to the new developments in Time–Dependent Density–Functional Theory. The hydrodynamical formulation expresses the long–wavelength xc potential of inhomogeneous systems in terms of the xc kernels of the homogeneous system, and connects them with generalized viscosity coefficients. Detailed analysis of the long–wavelength, small–frequency limit proves the presence of a discontinuity in the longitudinal component and relates the xc kernels to the elastic moduli. The high–frequency limit was also computed from Monte Carlo simulation results. We presented an approximate decoupling of the equation of motion for the current–current response function, which emphasizes the role of two–plasmon excitations. Based on this result, we formulated a simple self–consistent model for the long–wavelength plasmon dispersion in alkali metals, whose results are in qualitative agreement with experiment. The same approach has meanwhile been formally extended to the two–dimensional case [42,43], detailed numerical calculations in 2D are under way.

The second part of this work was based on the Diffusion Monte Carlo method, which is to date the method of choice for the study of static properties of the systems of interest here. We presented extensive results on 3D bosons and on 2D bosons and fermions, which allow to get a complete picture of the behaviour at varying coupling strength. Most results are in qualitative agreement with previous theoretical predictions, even if significant quantitative differences have been evidenced. Our results can be used both as a test for approximate theories, as discussed in Part II, and as an input for more complex many–body theories including e.g. dynamic effects, as was done in Part I. An explicit connection with available experimental data is discussed in Chapter 6.

The third part addresses many–body properties of systems of more direct experimental interest, i.e. various realizations of 2D electron gas in semiconductor heterostructures. Chapter 6 studies the electron–electron bilayer, whose phase diagram had been a subject of active debate. Diffusion Monte Carlo allows one to assess the zero–temperature phase diagram and equation of state of the corresponding idealized model, giving results in quantitative agreement with experiment. Chapter 7 addresses the tunneling properties of inhomogeneous Quantum Hall systems within an Independent Boson Model, which emphasizes the role of collective edge excitations. This leads to the prediction of a non–Fermi liquid behaviour for electron tunneling, with characteristic exponents which depend on both bulk filling factor and edge width.

In summary, this Thesis addressed various effects of correlations in the interacting electron fluid, both from almost exact numerical ground–state methods and via approximate dynamical theories, emphasizing the role of collective excitations.

Appendix A

Explicit evaluation of the low-k expansion

In this Appendix we explicitly compute the low-k expansion of the current-current response function. The basic definitions have been given in Sections 1.3 and 1.4, the equation-of-motion method has been presented in Section 1.3.3. The following calculations are performed in 3D, extension to 2D has been done in collaboration with R. Nifosì and is reported elsewhere [42]. The results below apply to both bosons and fermions. Section A.3 contains some simple relations which are useful to reproduce the following calculations.

A.1 High frequency limit

The third moment entering equation 1.27 is given by

$$\tilde{M}_3^{ij}(k) = \langle[[\mathbf{j}_\mathbf{k}^{(i)}, H], \mathbf{j}_{-\mathbf{k}}^{(j)}]\rangle = \tag{A.1}$$

$$= \frac{1}{m}\sum_\mathbf{q}\left[\frac{k^2}{2m}\left(\frac{\mathbf{k}^i\mathbf{k}^j}{2m} + \frac{2\mathbf{q}^i\mathbf{q}^j}{m}\right) + \frac{\mathbf{q}^i\mathbf{k}^j + \mathbf{q}^j\mathbf{k}^i}{m}\frac{(\mathbf{q}\cdot\mathbf{k})}{m}\right]n_\mathbf{q}$$

$$+ \frac{1}{V}\sum_\mathbf{q} v_q\langle\rho_\mathbf{q}\rho_{-\mathbf{q}}\rangle\left[\frac{\mathbf{q}^i}{m}\frac{(\mathbf{k}-\mathbf{q})^j}{m} + \frac{q^{d-1}}{|\mathbf{k}-\mathbf{q}|^{d-1}}\frac{(\mathbf{k}-\mathbf{q})^i(\mathbf{k}-\mathbf{q})^j}{m^2}\right] \tag{A.2}$$

where the behaviour $v_q \propto q^{1-d}$ has been used. By computing the angular average we get

$$\tilde{M}_3^{ij}(k) = \frac{k^2}{2m}\frac{^2\mathbf{k}^i\mathbf{k}^j}{2m}n + \frac{k^2\delta_{ij} + 2\mathbf{k}^i\mathbf{k}^j}{2m}\frac{4}{d}\delta^{ij}n\langle ke\rangle$$

$$+ \sum_\mathbf{q} v_q\frac{\mathbf{q}^i\mathbf{q}^j}{m^2}[S(\mathbf{q}+\mathbf{k}) - S(\mathbf{q})] \tag{A.3}$$

where $S(\mathbf{q}) = \langle \rho_\mathbf{q} \rho_{-\mathbf{q}} \rangle$, which gives directly equations (1.28–1.29). Expanding to small $\mathbf{k}$ we get

$$M_3^{L,T}(k) = \frac{k^2 n}{2m^2}\left(d_{L,T}\langle ke \rangle + e_{L,T}\langle pe \rangle\right) \tag{A.4}$$

where in 3D $d_L = 4$, $d_T = 4/3$, $e_L = 8/15$ and $e_T = -4/15$. The corresponding values for $f_{xc}^{L,T}$ (equations (1.37-1.38)) are obtained subtracting the ideal gas contribution and multiplying by the constant implied by equation (1.34). The situation is identical in two dimensions, where the coefficients are $d_L = 6$, $d_T = 2$, $e_L = 5/4$ and $e_T = -1/4$.

A.2 Two–pair contributions at $k = 0$ and finite ω

In this section we explicitly evaluate the two–pair contributions to $\mathbf{Im}\,\chi_{ij}$. Equation (1.27) can be written as

$$
\begin{aligned}
\omega^2 \tilde{\chi}_{ij}(\mathbf{k},\omega) \;=\;& \tilde{M}_3^{ij}(k) \\
& -\ll \left[j_\mathbf{k}^{(i)}, KE\right];\ \left[j_{-\mathbf{k}}^{(j)}, KE\right] \gg \\
& -\ll \left[j_\mathbf{k}^{(i)}, KE\right];\ \left[j_{-\mathbf{k}}^{(j)}, PE'\right] \gg \\
& -\ll \left[j_\mathbf{k}^{(i)}, PE\right];\ \left[j_{-\mathbf{k}}^{(j)}, KE\right] \gg \\
& -\ll \left[j_\mathbf{k}^{(i)}, PE'\right];\ \left[j_{-\mathbf{k}}^{(j)}, PE'\right] \gg
\end{aligned}
\tag{A.5}
$$

where $\tilde{M}_3^{ij}(k)$ was evaluated above. Only the last term in the previous equation is already explicitly second order in the interaction potential. At the same time, since each commutator with the kinetic energy gives a factor of k, in order to get the full expansion to order k^2 terms with up to two commutators with the kinetic energy are needed. The relevant ones are

$$
\begin{aligned}
\mathbf{Im}\,\omega^2 \tilde{\chi}_{ij}(\mathbf{k},\omega) \;=\;& \frac{1}{\omega^2}\mathbf{Im} \ll \left[\left[j_\mathbf{k}^{(i)}, KE\right], PE'\right];\ \left[\left[j_{-\mathbf{k}}^{(j)}, KE\right], PE'\right] \gg \\
& +\frac{1}{\omega^2}\mathbf{Im} \ll \left[\left[j_\mathbf{k}^{(i)}, KE\right], PE\right];\ \left[\left[j_{-\mathbf{k}}^{(j)}, PE'\right], KE\right] \gg \\
& +\frac{1}{\omega^2}\mathbf{Im} \ll \left[\left[j_\mathbf{k}^{(i)}, PE\right], KE\right];\ \left[\left[j_{-\mathbf{k}}^{(j)}, KE\right], PE\right] \gg \\
& -\mathbf{Im} \ll \left[j_\mathbf{k}^{(i)}, PE'\right];\ \left[j_{-\mathbf{k}}^{(j)}, PE'\right] \gg
\end{aligned}
\tag{A.6}
$$

These contributions are evaluated in the remaining part of this Appendix, using the decoupling (2.2). The second and the third, being very similar, will be treated together.

A.2.1   PP term

$$PP_{ij} = \mathbf{Im} \ll \left[\mathbf{j}_{\mathbf{k}}^{(i)}, PE'\right]; \left[\mathbf{j}_{-\mathbf{k}}^{(j)}, PE'\right] \gg \tag{A.7}$$

This term is explicitly of order 0 in k. Careful evaluation of the low-k expansion in an uniform and isotropic system shows that the coefficient of the leading k^0 term vanishes, in agreement with the basic fact that the plasmon is the only excitation to leading order in k.

The complete expansion of PP_{ij} is

$$PP_{ij} = \sum_{\mathbf{q},\mathbf{q'}} v_q v_{q'} \frac{q^i q'^j}{m^2} \mathbf{Im} \ll \rho_{\mathbf{k}-\mathbf{q}}\rho_{\mathbf{q}}; \ \rho_{-\mathbf{k}-\mathbf{q'}}\rho_{\mathbf{q'}} \gg \tag{A.8}$$

and, after performing the RPA decoupling,

$$PP_{ij} = -\frac{1}{m^2} \sum_{\mathbf{q}} v_{\mathbf{q}+\mathbf{k}/2} \left(\mathbf{q}+\frac{\mathbf{k}}{2}\right)^i \left[v_{\mathbf{q}-\mathbf{k}/2}\left(\mathbf{q}-\frac{\mathbf{k}}{2}\right)^j - v_{\mathbf{q}+\mathbf{k}/2}\left(\mathbf{q}+\frac{\mathbf{k}}{2}\right)^j\right]$$

$$\times \int_0^\omega \frac{d\omega'}{\pi} \mathbf{Im} \ll \rho_{\mathbf{q}+\mathbf{k}/2}; \ \rho_{-\mathbf{q}-\mathbf{k}/2} \gg \mathbf{Im} \ll \rho_{\mathbf{q}-\mathbf{k}/2}; \ \rho_{-\mathbf{q}+\mathbf{k}/2} \gg \tag{A.9}$$

Low-k expansion and angular averaging lead to

$$PP_{L,T} = c_{L,T}^{PP} \frac{k^2}{m^2} \int_0^\omega \frac{d\omega'}{\pi} \sum_{\mathbf{q}} v_q^2 \frac{q^2}{\omega'^2} \frac{q^2}{(\omega-\omega')^2} \mathbf{Im}\,\chi_L(\mathbf{q},\omega') \mathbf{Im}\,\chi_L(\mathbf{q},\omega-\omega') \tag{A.10}$$

where (in $d=3$) $c_L^{PP} = 7/30$ and $c_T^{PP} = 4/30$ for the longitudinal and transverse component respectively.

A.2.2   KPKP term

The term $\ll \left[\mathbf{j}_{\mathbf{k}}^{(i)}, KE\right]; \left[\mathbf{j}_{-\mathbf{k}}^{(j)}, KE\right] \gg$ contains only single-pair terms, and indeed has contributions to zeroth order in the interaction potential. It is explicitly of order k^2, therefore further commutators with the kinetic energy give no contribution to order k^2. The relevant term is

$$KPKP_{ij} = \ll \left[\left[\mathbf{j}_{\mathbf{k}}^{(i)}, KE\right], PE'\right]; \left[\left[\mathbf{j}_{-\mathbf{k}}^{(j)}, KE\right], PE'\right] \gg =$$

$$= \sum_{\mathbf{hh'}} v_h v_{h'} \left[-\frac{h^i}{m}k^l - \frac{\mathbf{h}\cdot\mathbf{k}}{m}\delta^{il}\right]\left[\frac{h'^j}{m}k^m + \frac{\mathbf{h'}\cdot\mathbf{k}}{m}\delta^{jm}\right]$$

$$\times \ll \mathbf{j}_{\mathbf{k}+\mathbf{h}}^{(l)}\rho_{-\mathbf{h}}; \ \mathbf{j}_{-\mathbf{k}+\mathbf{h'}}^{(m)}\rho_{-\mathbf{h'}} \gg \tag{A.11}$$

where only two-pair terms have been included. Upon low-k expansion and angular averaging we get

$$KPKP_{L,T} = -\sum_{\mathbf{h}} \int_0^\omega \frac{d\omega'}{\pi} v_h^2 \frac{h^4 k^2}{m^2}\left[c_{L,T}^{KP}\frac{1}{(\omega-\omega')^2}\mathbf{Im}\,\chi_T(h,\omega')\mathbf{Im}\,\chi_L(h,\omega-\omega')\right.$$

$$\left.+d_{L,T}^{KP}\frac{\omega^2}{\omega'^2(\omega-\omega')^2}\mathbf{Im}\,\chi_L(h,\omega')\mathbf{Im}\,\chi_L(h,\omega-\omega')\right] \tag{A.12}$$

with $c_L^{KP} = 8/15$, $d_L^{KP} = 2/5$, $c_T^{KP} = 2/5$, and $d_T^{KP} = 2/15$ in 3D.

A.2.3 KPPK term

$$KPPK_{ij} = \ll \left[[\mathbf{j}_{\mathbf{k}}^{(i)}, KE], PE' \right]; \left[[\mathbf{j}_{-\mathbf{k}}^{(j)}, PE'], KE \right] \gg \qquad (A.13)$$

Analogous considerations show that the only relevant contribution is

$$KPPK_{ij} = -\sum_{\mathbf{h}} \int_0^\omega \frac{d\omega'}{\pi} v_h^2 2 \left[h^2 \frac{\mathbf{h} \cdot \mathbf{k}}{m^2} \mathbf{h}^i \mathbf{k}^j - 2 \frac{(\mathbf{h} \cdot \mathbf{k})^2}{m^2} \mathbf{h}^i \mathbf{h}^j \right] \frac{\omega}{\omega'(\omega - \omega')^2}$$
$$\times \mathbf{Im}\,\chi_L(h, \omega') \mathbf{Im}\,\chi_L(h, \omega - \omega') \qquad (A.14)$$

Inclusion of the symmetric term $\ll \left[[\mathbf{j}_{-\mathbf{k}}^{(i)}, PE'], KE \right]; \left[[\mathbf{j}_{\mathbf{k}}^{(j)}, KE], PE' \right] \gg$ leads to

$$[\dots] \to \left[h^2 (\mathbf{h} \cdot \mathbf{k})(\mathbf{h}^i \mathbf{k}^j + \mathbf{k}^i \mathbf{h}^j) - 4 (\mathbf{h} \cdot \mathbf{k})^2 \mathbf{h}^i \mathbf{h}^j \right] \qquad (A.15)$$

Finally,

$$KPPK_{L,T} = -\sum_{\mathbf{h}} \int_0^\omega \frac{d\omega'}{\pi} v_h^2 c_{L,T}^{KPPK} \frac{h^4 k^2}{m^2} \frac{\omega^2}{\omega'^2 (\omega - \omega')^2} \mathbf{Im}\,\chi_L(h, \omega') \mathbf{Im}\,\chi_L(h, \omega - \omega')$$

$$(A.16)$$

with $c_L^{KPPK} = 2/15$, $c_T^{KPPK} = 4/15$. Collecting the previous results we get

$$\mathbf{Im}\,\tilde{\chi}(k, \omega) = -\frac{k^2 n^2}{\omega^2 m^2} \int_0^\omega \frac{d\omega'}{\pi} \int \frac{d^3 q}{(2\pi)^3 n^2} v_q^2 \frac{q^2}{(\omega - \omega')^2} \mathbf{Im}\,\chi_L(q, \omega - \omega')$$

$$\times \left[a_{L,T} \frac{q^2}{\omega'^2} \mathbf{Im}\,\chi_L(q, \omega') + b_{L,T} \frac{q^2}{\omega^2} \mathbf{Im}\,\chi_T(q, \omega') \right] \qquad (A.17)$$

with $a_L = 23/30$, $a_T = 8/15$, $b_L = 8/15$ and $b_T = 2/5$, which leads directly to equation (2.3).

A.2.4 Comparison with the NSSS model

A related model was introduced a few years ago by Neilson *et al* [45], within the memory function formalism. Transverse currents being absent in their formalism, they obtained

$$\mathbf{Im}\,G(k, \omega) = \frac{1}{\pi n m \omega_{pl}^2 k^2} \int_0^\omega d\omega' \sum_{\mathbf{q}} \bar{v}_{\mathbf{q}} (\mathbf{q} \cdot \mathbf{k}) \mathbf{Im}\,\chi(\mathbf{q}, \omega') \qquad (A.18)$$

$$\times \mathbf{Im}\,\chi(\mathbf{k} - \mathbf{q}, \omega - \omega') \left[\bar{v}_{\mathbf{q}} (\mathbf{q} \cdot \mathbf{k}) + \bar{v}_{\mathbf{k}-\mathbf{q}} (k^2 - \mathbf{q} \cdot \mathbf{k}) \right] .$$

where $\bar{v}_k$ is a statically screened potential of the STLS type [56]. The coupled–mode structure of the dynamic local field factor is again evident from equation (A.18), although the transverse response function entering equation (A.17) is missing.

By taking $\bar{v}_k = v_k$ we obtain a result identical with the term discussed in Sec. A.2.1 alone (equation (A.9)), with no transverse currents.

While the NSSS result is obviously superior in being valid at finite k, it does not consistently include all $O(k^2)$ terms. From a perturbational point of view, the full expression gives the correct perturbational limit derived by Glick and Long [48] after inclusion of exchange through equation (2.4), whereas the NSSS approach gives a result smaller by a factor of 7/23.

A.3   Some useful relations

This section contains some basic relations useful to verify the above equations.

Current and density response functions

The different response functions can be related as

$$\mathrm{Im} \ll \mathbf{j_q}; \rho_{-h} \gg = -\mathrm{Im} \ll \rho_{-h}; \mathbf{j_q} \gg = \delta_{\mathbf{qh}} \frac{\mathbf{q}}{\omega} \mathrm{Im}\, \chi_L(q, \omega) \tag{A.19}$$

$$\chi_{\rho\rho}(k, \omega) = \frac{nk^2}{m\omega^2} + \frac{k^2}{\omega^2}\chi_L(k, \omega) \tag{A.20}$$

Basic commutators

$$[\mathbf{j_k}, \rho_h] \;\; = \;\; -\frac{\mathbf{h}}{m}\rho_{h+k} \tag{A.21}$$

$$[\rho_k, H] \;\; = \;\; \mathbf{k} \cdot \mathbf{j_k} \tag{A.22}$$

$$[\mathbf{j_k}, PE] \;\; = \;\; \sum_{\mathbf{q}} v_q \frac{\mathbf{q}}{m}\rho_{\mathbf{k-q}}\rho_{\mathbf{q}} \tag{A.23}$$

$$[\mathbf{j_k}, KE] \;\; = \;\; \sum_{\mathbf{q}} \frac{\mathbf{q}}{m}\frac{\mathbf{q} \cdot \mathbf{k}}{m} c^{\dagger}_{\mathbf{q-k/2}} c_{\mathbf{q+k/2}} \tag{A.24}$$

$$[[\mathbf{j_k}, KE], \rho_h] \;\; = \;\; -\frac{\mathbf{h}}{m}\mathbf{k} \cdot \mathbf{j_{k+h}} - \frac{\mathbf{h} \cdot \mathbf{k}}{m}\mathbf{j_{k+h}} \tag{A.25}$$

$$[[\mathbf{j_k}, KE], PE] \;\; = \;\; \sum_{\mathbf{h}} v_h \left[-\frac{\mathbf{h}}{m}\mathbf{k} \cdot \mathbf{j_{k+h}}\rho_{-h} - \frac{\mathbf{h} \cdot \mathbf{k}}{m}\mathbf{j_{k+h}}\rho_{-h} + \frac{\mathbf{k} \cdot \mathbf{h}}{m}\frac{\mathbf{h}}{m}\rho_k \right]$$

$$[[\mathbf{j_k}, PE], KE] \;\; = \;\; \sum_{\mathbf{q}} v_q \frac{\mathbf{q}}{m} [\rho_{\mathbf{k-q}}\mathbf{q} \cdot \mathbf{j_q} + (\mathbf{k} - \mathbf{q}) \cdot \mathbf{j_{k-q}}\rho_{\mathbf{q}}] = \tag{A.26}$$

$$= \sum_{\mathbf{q}} \left[v_{\mathbf{q+k/2}}\frac{\mathbf{q+k/2}}{m} - v_{\mathbf{q-k/2}}\frac{\mathbf{q-k/2}}{m} \right] \rho_{\mathbf{k/2-q}} \left(\mathbf{q} + \frac{\mathbf{k}}{2} \right) \cdot \mathbf{j_{q+k/2}} =$$

$$= \sum_{\mathbf{q}} v_q \left(\frac{\mathbf{k}}{m} + (1 - d)\frac{\mathbf{k} \cdot \mathbf{q}}{mq^2}\mathbf{q} \right) \rho_{-q}\mathbf{q} \cdot \mathbf{j_q} + O(k^2) \tag{A.27}$$

Angular averages

$$\langle (\mathbf{k} \cdot \mathbf{q})^2 \rangle = \frac{1}{d} k^2 q^2 \tag{A.28}$$

$$\langle (\mathbf{k} \cdot \mathbf{q})^4 \rangle = \begin{cases} \frac{1}{5} k^4 q^4 & d = 3 \\ \frac{3}{8} k^4 q^4 & d = 2 \end{cases} \tag{A.29}$$

Decomposition in longitudinal and transverse components

$$A_{ij} = \delta_{ij} A_T + \hat{k}_i \hat{k}_j \left(A_L - A_T \right) \tag{A.30}$$

$$A_L = \frac{\mathbf{k}^i \mathbf{k}^j}{k^2} A_{ij} \tag{A.31}$$

$$(d-1) A_T = \delta_{ij} A_{ij} - A_L . \tag{A.32}$$

Appendix B

Transverse response function of noninteracting fermions

The current–current response function for the noninteracting Fermi gas is given by

$$
\chi^0_{ij}(k, \omega) = \frac{1}{m^2} \sum_{\mathbf{q}} q^i q^j \frac{n^F_{\mathbf{q}+\mathbf{k}/2} - n^F_{\mathbf{q}-\mathbf{k}/2}}{\omega - (\varepsilon_{\mathbf{q}-\mathbf{k}/2} - \varepsilon_{\mathbf{q}+\mathbf{k}/2}) + i\eta} = \tag{B.1}
$$

$$
= \frac{1}{m^2} \sum_{\mathbf{q}} n^F_{\mathbf{q}} \left(\mathbf{q} + \frac{1}{2}\mathbf{k}\right)^i \left(\mathbf{q} + \frac{1}{2}\mathbf{k}\right)^j
$$

$$
\times \left[\frac{1}{\omega + i\eta + \varepsilon_{\mathbf{q}} - \varepsilon_{\mathbf{q}+\mathbf{k}}} - \frac{1}{\omega + i\eta - \varepsilon_{\mathbf{q}} + \varepsilon_{\mathbf{q}+\mathbf{k}}} \right] \tag{B.2}
$$

While the longitudinal component can be easily obtained from the usual Lindhard function, the transverse needs to be explicitly computed. It is sufficient to evaluate

$$
A_{ij}(k, \omega) = -\frac{\pi}{m^2} \sum_{\mathbf{q}} n^F_{\mathbf{q}} \left(\mathbf{q} + \frac{1}{2}\mathbf{k}\right)^i \left(\mathbf{q} + \frac{1}{2}\mathbf{k}\right)^j \delta\left(\omega + \varepsilon_{\mathbf{q}} - \varepsilon_{\mathbf{q}+\mathbf{k}}\right) \tag{B.3}
$$

since $\mathbf{Im}\,\chi_{ij}(k, \omega) = A_{ij}(k, \omega) - A_{ij}(k, -\omega)$. Angular integration shows that A is nonvanishing only for $|\omega - \varepsilon_k| < kk_F/m$, where

$$
A(k, \omega) = -\frac{1}{2\pi m^2} \int_{\frac{|\omega - \varepsilon_k|}{k/m}}^{k_F} q^2 dq \frac{m}{kq} \begin{cases} (qx + k/2)^2 & \text{(Long.)} \\ q^2(1 - x^2)/2 & \text{(Trans.)} \end{cases} \tag{B.4}
$$

where $x = \cos\theta_{\mathbf{kq}} = (\omega - \varepsilon_k)/(kq/m)$. This equation shows that at large k the longitudinal part is larger than the transverse one by a factor of order k^2/k_F^2. Explicit integration leads to

$$
\mathbf{Im}\,\chi^0_T(k, \omega) = -\frac{\omega}{16\pi k^3}\left(4k^2 k_F^2 - k^4 - 4m^2\omega^2\right) \tag{B.5}
$$

for $0 < \omega < (2kk_F - k^2)/2m$ and

$$\mathbf{Im}\,\chi_T^0(k,\omega) = -\frac{1}{256\pi\,m\,k^5}\left(k^4 - 4\,k^2\,k_F^2 - 4\,k^2\,m\,\omega + 4\,m^2\,\omega^2\right)^2 \qquad (B.6)$$

for $|2kk_F - k^2|/2m < \omega < (2kk_F + k^2)/2m$. The imaginary part of the transverse response function is zero elsewhere.

An additional check of the validity of the present result can be obtained by explicit evaluation of the first frequency moment sum rule, which leads to

$$-\frac{2}{\pi}\int_0^\infty \omega\mathbf{Im}\,\chi_T^0(k,\omega)d\omega = \frac{4}{5}n\frac{k^2}{2m^2}\frac{k_F^2}{2m} \qquad (B.7)$$

in agreement with equation (A.4), where – to zero order in v_k – $\langle ke\rangle = \frac{3}{5}k_F^2/2m$ and $\langle pe\rangle = 0$.

The Kramers–Kronig transform of the result in equations (B.5) and (B.6) yields

$$\mathbf{Re}\,\chi_T^0(k,\omega) = \frac{k_F}{96\pi^2\,m\,k^2}\left(3\,k^4 - 20\,k^2\,k_F^2 + 36\,m^2\,\omega^2\right) \qquad (B.8)$$

$$+\frac{1}{256\pi^2\,m\,k^5}\left(k^4 - 4k^2k_F^2 + 4k^2m\omega + 4m^2\omega^2\right)^2 \ln\left|\frac{\omega + k^2/2m - kk_F/m}{\omega + k^2/2m + kk_F/m}\right|$$

$$+\frac{1}{256\pi^2\,m\,k^5}\left(k^4 - 4k^2k_F^2 - 4k^2m\omega + 4m^2\omega^2\right)^2 \ln\left|\frac{\omega - k^2/2m + kk_F/m}{\omega - k^2/2m - kk_F/m}\right|.$$

We now turn to examine some limiting behaviours. At long wavelength

$$\chi_T^0(k\to0,\omega) = \frac{4n}{5m\omega^2}\frac{k^2}{2m}\frac{k_F^2}{2m} + \frac{48}{35}\left(\frac{k_F^2}{2m}\right)^2\left(\frac{k^2}{2m}\right)^2\frac{n}{m\omega^4}, \qquad (B.9)$$

at zero frequency

$$\chi_T^0(k,\omega=0) = -\frac{5n}{8m} + \frac{k^2k_F}{32m\pi^2} + \frac{1}{128km\pi^2}\left(4k_F^2 - k^2\right)^2\ln\left|\frac{k - 2k_F}{k + 2k_F}\right| \qquad (B.10)$$

and the static long wavelength limit is

$$\chi_T^0(k\to0,\omega=0) = -\frac{n}{m} + \frac{k_Fk^2}{12m\pi^2} + \cdots \qquad (B.11)$$

in agreement with Ref. [31]. The $k\to0$, $\omega\to0$ limit clearly depends on the order in which the limits are taken, as in the longitudinal case.

Acknowledgements

It is difficult for me to express in a few words my indebtedness to Prof. Mario Tosi, not only for the scientific advise and for the attention he dedicated to this work but also for encouraging me to travel and interact with people from other universities.

Large part of this thesis relies on the months I spent at the University of Missouri at Columbia, under the guidance of Professor Giovanni Vignale. Giovanni has been able to teach me *a lot* of different things in a short period, and my main regret is to have been unable to follow more quickly his fast–developing ideas.

I thank Professor Gaetano Senatore, from the University of Trieste, for many stimulating discussions, including some work in progress on Monte Carlo simulations of defects in solid Helium^4, as well as for the interesting collaboration on the electron–electron bilayer.

Special thanks are due to Saverio Moroni, also from Trieste, who taught me the Diffusion Monte Carlo method and was always ready to come up with the right explanation any time something strange happened during the simulation. I am indebted with Saverio also for allowing me to use his DMC computer code. The numerical part of this Thesis benefited also from many discussions with Francesco Rapisarda, from the University of Linz.

I wish to thank Helga Böhm, from Linz as well, with whom I started the work presented in Chapter 2. The successive developments of this work were stimulated by suggestions of Giovanni Vignale, and benefited from a recent collaboration with Riccardo Nifosì on the two–dimensional case.

I thank Fabio Pistolesi, who was always ready to discuss serisouly *any* topic, usually leading to a much deeper understanding of the issue. I also thank Valentina Tozzini, with whom I shared the office, for a large number of useful suggestions. And she wasn't even upset because of my hours–long phone calls with Gaetano Senatore and Giovanni Vignale.

Finally, I wish to acknowledge Fabio Beltram, Franco Buda, Marilú Chiofalo, Irene D'Amico, Silvano De Franceschi, Paolo Giannozzi, Giuseppe La Rocca, Anna Minguzzi and Carsten Ullrich, for fruitful discussions.

Bibliography

[1] P. Hohenberg and W. Kohn, Phys. Rev. **136**, B864 (1964).

[2] W. Kohn and L. Sham, Phys. Rev. **140**, A1133 (1965).

[3] R. G. Parr and W. Yang, *Density-Functional Theory of Atoms and Molecules* (Oxford University Press, New York, 1989).

[4] R. M. Dreizler and E. K. U. Gross, *Density functional theory, an approach to the quantum many–body problem* (Springer, Berlin, 1990).

[5] E. Runge and E. K. U. Gross, Phys. Rev. Lett. **52**, 997 (1984).

[6] E. K. U. Gross, J. F. Dobson, and M. Petersilka, in *Density Functional Theory, Topics in Current Chemistry*, edited by R. F. Nalewajski (Springer, Berlin, 1996).

[7] A. Zangwill and P. Soven, Phys. Rev. Lett. **45**, 204 (1980).

[8] A. Zangwill and P. Soven, Phys. Rev. B **24**, 4121 (1981).

[9] W. Ekardt, Phys. Rev. Lett. **52**, 1925 (1984).

[10] W. Ekardt, Phys. Rev. B **31**, 6360 (1985).

[11] E. K. U. Gross and W. Kohn, Phys. Rev. Lett. **55**, 2850 (1985), erratum in Phys. Rev. Lett. **57**, 923 (1986).

[12] E. K. U. Gross and W. Kohn, Phys. Rev. Lett. **57**, 923 (1986).

[13] N. Iwamoto and E. K. U. Gross, Phys. Rev. B **35**, 3003 (1987).

[14] J. Dobson, Phys. Rev. Lett. **73**, 2244 (1994).

[15] J. Dobson, in *Density Functional Theory, NATO ASI*, edited by E. K. U. Gross and R. M. Dreizler (Plenum, New York, 1994), p. 393.

[16] G. Vignale, Phys. Rev. Lett. **74**, 3233 (1995).

[17] G. Vignale, Phys. Lett. A **209**, 206 (1995).

[18] G. Vignale and W. Kohn, Phys. Rev. Lett. **77**, 2037 (1996).

[19] G. Vignale and W. Kohn, in *Electronic Density Functional Theory*, edited by J. Dobson, M. P. Das, and G. Vignale (Plenum Press, New York, 1997).

[20] B. Dabrowski, Phys. Rev. B **34**, 4989 (1986).

[21] A. vom Felde, J. Sprösser-Prou, and J. Fink, Phys. Rev. B **40**, 10181 (1989).

[22] J. Sprösser-Prou, A. vom Felde, and J. Fink, Phys. Rev. B **40**, 5799 (1989).

[23] B. C. Larson, J. Z. Tishler, E. D. Isaacs, P. Zschack, A. Fleszar, and A. G. Eguiluz, Phys. Rev. Lett. **77**, 1346 (1996).

[24] L. D. Landau and E. Lifshitz, *Mechanics of Fluids*, Vol. 6 of *Course of theoretical Physics*, 2nd ed. (Pergamon Press, Oxford, 1987).

[25] L. D. Landau and E. Lifshitz, *Theory of Elasticity*, Vol. 7 of *Course of theoretical Physics*, 3rd ed. (Pergamon Press, Oxford, 1986).

[26] J. F. Dobson, M. J. Bünner, and E. K. U. Gross, Phys. Rev. Lett. **79**, 1905 (1997).

[27] D. Pines and P. Nozières, *The Theory of Quantum Liquids* (W. A. Benjamin, New York, 1966), Vol. 1.

[28] J. Lindhard, Mat.-Fys. Medd.-K. Dan. Vidensk. Selsk. **28**, (1954).

[29] P. Nozières, *The Theory of Interacting Fermi Systems* (W. A. Benjamin, New York, 1964).

[30] K. S. Singwi and M. P. Tosi, in *Solid State Physics*, edited by H. Ehrenreich, F. Seitz, and D. Turnbull (Academic, New York, 1981), Vol. 36, p. 177.

[31] G. Vignale, M. Rasolt, and D. J. W. Geldart, Phys. Rev. B **37**, 2502 (1988).

[32] D. M. Ceperley and B. J. Alder, Phys. Rev. Lett. **45**, 566 (1980).

[33] S. H. Vosko, L. Wilk, and M. Nusair, Can. J. Phys. **58**, 1200 (1980).

[34] T. M. Rice, Ann. Phys. (N.Y.) **31**, 100 (1965).

[35] L. Hedin, Phys. Rev. **139A**, 796 (1965).

[36] H. Yasuhara and Y. Ousaka, Int. J. Mod. Phys. B **6**, 3089 (1992).

[37] Y. Kwon, D. M. Ceperley, and R. M. Martin, Phys. Rev. B **50**, 1684 (1994).

[38] E. Lipparini, S. Stringari, and K. Takayanagi, J. Phys.: Condens. Matter **6**, 2025 (1994).

[39] H. M. Böhm, S. Conti, and M. P. Tosi, J. Phys.: Condens. Matter **8**, 781 (1996).

[40] S. Conti, H. M. Böhm, and M. P. Tosi, Phys. Stat. Sol (b) **193**, K11 (1996).

[41] S. Conti, R. Nifosì, and M. P. Tosi, J. Phys.: Condens. Matter **9**, L475 (1997).

[42] R. Nifosí, *Potenziali dinamici di scambio e correlazione in fluidi elettronici*, Tesi di Laurea, Università di Pisa, 1997.

[43] R. Nifosì, S. Conti, and M. P. Tosi, Physica E **1**, 188 (1997).

[44] M. Hasegawa and M. Watabe, J. Phys. Soc. Jpn. **27**, 1393 (1969).

[45] D. Neilson, L. Swierkowski, A. Sjölander, and J. Szymanski, Phys. Rev. B **44**, 6291 (1991).

[46] D. F. DuBois, Ann. Phys. (NY) **8**, 24 (1959).

[47] D. F. DuBois and M. G. Kivelson, Phys. Rev. B **186**, 409 (1969).

[48] A. J. Glick and W. F. Long, Phys. Rev. B **4**, 3455 (1971).

[49] B. W. Ninham, C. J. Powell, and N. Swanson, Phys. Rev. **145**, 209 (1966).

[50] W. Gasser, Zs. Phys. B **57**, 15 (1984).

[51] M. E. Bachlechner, W. Macke, H. M. Miesenböck, and A. Schinner, Physica B **168**, 104 (1991).

[52] W. Gasser, Physica B **183**, 217 (1992).

[53] M. E. Bachlechner, H. M. Böhm, and A. Schinner, Phys. Lett. A **178**, 186 (1993).

[54] S. Conti, M. L. Chiofalo, and M. P. Tosi, J. Phys.: Condens. Matter **6**, 8795 (1994).

[55] M. L. Chiofalo, S. Conti, S. Stringari, and M. P. Tosi, J. Phys.: Condens. Matter **7**, L85 (1995).

[56] K. S. Singwi, M. P. Tosi, R. H. Land, and A. Sjölander, Phys. Rev. **176**, 589 (1968).

[57] P. Vashishta and K. S. Singwi, Phys. Rev. B **6**, 875 (1972).

[58] S. Moroni, D. Ceperley, and G. Senatore, Phys. Rev. Lett. **75**, 689 (1995).

[59] K. N. Pathak and P. Vashishta, Phys. Rev. B **7**, 3649 (1973).

[60] K. Utsumi and S. Ichimaru, Phys. Rev. A **26**, 603 (1982).

[61] S. Ichimaru, Rev. Mod. Phys. **54**, 1017 (1982).

[62] J. L. Parish, Phys. Rev. B **42**, 10940 (1990).

[63] L. Serra, F. Garcias, M. Barranco, N. Barberan, and J. Navarro, Phys. Rev. B **44**, 1492 (1991).

[64] V. A. Ignatchenko and Y. I. Mankov, J. Phys.: Condens. Matter **3**, 5837 (1991).

[65] G. Kalman, K. Kempa, and M. Minella, Phys. Rev. B **43**, 14238 (1991).

[66] M. Taut and K. Sturm, Sol. State Commun. **82**, 295 (1992).

[67] F. Aryasetiawan and K. Karlsson, Phys. Rev. Lett. **73**, 1679 (1994).

[68] A. A. Quong and A. G. Eguiluz, Phys. Rev. Lett. **70**, 3955 (1993).

[69] A. Fleszar, R. Stumpf, and A. G. Eguiluz, preprint (unpublished).

[70] H. Suehiro, Y. Ousaka, and H. Yasuhara, J. Phys. C **18**, 6007 (1985).

[71] S. Moroni, S. Conti, and M. P. Tosi, in *Physics of Strongly Coupled Plasmas*, edited by W. D. Kraeft and M. Schlanges (World Scientific, Singapore, 1995), p. 419.

[72] D. M. Ceperley, Phys. Rev. B **18**, 3126 (1978).

[73] D. M. Ceperley and M. H. Kalos, in *Monte Carlo Methods in Statistical Physics*, edited by K. Binder (Springer, Berlin, 1979).

[74] P. J. Reynolds, D. M. Ceperley, B. J. Alder, and W. A. Lester, J. Chem. Phys. **77**, 5593 (1982).

[75] Y. Kwon, D. M. Ceperley, and R. M. Martin, Phys. Rev. B **48**, 12037 (1993).

[76] J. Casulleras and J. Boronat, Phys. Rev. B **52**, 3654 (1995).

[77] Y. Kwon, D. M. Ceperley, and R. M. Martin, Phys. Rev. B **53**, 7376 (1996).

[78] C. J. Umrigar, M. P. Nightingale, and K. J. Runge, J. Chem. Phys. **99**, 2865 (1993).

[79] V. Natoli and D. Ceperley, J. Comp. Phys. **117**, 171 (1995).

[80] S. Moroni, D. M. Ceperley, and G. Senatore, Phys. Rev. Lett. **69**, 1837 (1992).

[81] G. Sugiyama, C. Bowen, and B. J. Alder, Phys. Rev. B **46**, 13042 (1992).

[82] F. Y. Wu and E. Feenberg, Phys. Rev. **128**, 943 (1962).

[83] R. Monnier, Phys. Rev. A **5**, 814 (1972).

[84] J. P. Hansen and R. Mazighi, Phys. Rev. A **18**, 1282 (1978).

[85] M. R. Schafroth, Phys. Rev. **96**, 1149 (1954).

[86] A. S. Alexandrov and N. F. Mott, Supercond. Sci. Technol. **6**, 215 (1993).

[87] B. W. Ninham, Phys. Lett. **4**, 278 (1963).

[88] Schramm and K. Langanke, Astrophys. J. **397**, 579 (1992).

[89] H. M. Müller and K. Langanke, Phys. Rev. C **49**, 524 (1994).

[90] L. L. Foldy, Phys. Rev. **124**, 649 (1961).

[91] K. A. Brückner, Phys. Rev. **156**, 204 (1967).

[92] S. K. Ma and C. W. Woo, Phys. Rev. **159**, 165 (1967).

[93] D. K. Lee, Phys. Rev. **187**, 326 (1969).

[94] D. K. Lee and F. H. Ree, Phys. Rev. A **5**, 814 (1972).

[95] M. Saarela, Phys. Rev. B **29**, 191 (1984).

[96] V. Apaja, J. Halinen, V. Halonen, E. Krotcheck, and M. Saarela, Phys. Rev. B **55**, 12925 (1997).

[97] S. R. Hore and N. E. Frankel, Phys. Rev. B **12**, 2619 (1975).

[98] A. Caparica and O. Hipólito, Phys. Rev. A **26**, 2832 (1982).

[99] A. Gold, Z. Phys. B **89**, 1 (1992).

[100] S. Moroni, S. Conti, and M. P. Tosi, Phys. Rev. B **53**, 9688 (1996).

[101] D. M. Ceperley and B. J. Alder, J. Phys. Colloque **C7**, 295 (1980).

[102] H. R. Glyde, G. H. Keech, R. Mazighi, and J. P. Hansen, Phys. Lett. A **58**, 226 (1978).

[103] W. J. Carr, Jr., Phys. Rev. **122**, 1437 (1961).

[104] A. Holas, in *Strongly Coupled Plasma Physics*, edited by F. J. Rogers and H. E. DeWitt (Plenum Press, New York, 1986), p. 463.

[105] S. Moroni (unpublished).

[106] J. C. Kimball, J. Phys. A **8**, 1513 (1975).

[107] M. L. Chiofalo, S. Conti, and M. P. Tosi, J. Phys.: Condens. Matter **8**, 1921 (1996).

[108] S. Giorgini and S. Stringari (unpublished).

[109] J. Boronat, J. Casulleras, F. Dalfovo, S. Moroni, and S. Stringari, Phys. Rev. B **52**, 1236 (1995).

[110] M. L. Chiofalo, S. Conti, and M. P. Tosi, Mod. Phys. Lett. B **8**, 1207 (1994).

[111] M. L. Ristig and J. W. Clark, Phys. Rev. B **40**, 4355 (1989).

[112] N. Hugenholtz and D. Pines, Phys. Rev. **116**, 489 (1959).

[113] E. Wigner, Trans. Fraday Soc. **34**, 678 (1938).

[114] F. Rapisarda and G. Senatore, Aust. J. Phys. **49**, 161 (1996).

[115] B. Tanatar and D. M. Ceperley, Phys. Rev. B **39**, 5005 (1989).

[116] R. M. May, Phys. Rev. **115**, 254 (1959).

[117] C. I. Um, W. H. Kahng, E. S. Yim, and T. F. George, Phys. Rev. B **41**, 259 (1990).

[118] R. K. Moudgil, P. K. Ahluwalia, K. Tankeshwar, and K. N. Pathak, Phys. Rev. B **55**, 544 (1997).

[119] H. K. Sim, R. Tao, and F. Y. Wu, Phys. Rev. B **34**, 7123 (1986).

[120] A. K. Rajagopal and J. C. Kimball, Phys. Rev. B **15**, 2819 (1977).

[121] S. Moroni, S. Fantoni, and G. Senatore, Phys. Rev. B **52**, 13547 (1995).

[122] W. R. Magro and D. M. Ceperley, Phys. Rev. Lett. **73**, 826 (1994).

[123] G. E. Santoro and G. F. Giuliani, Phys. Rev. B **39**, 12818 (1989).

[124] S. Moroni, G. Senatore, and S. Fantoni, Phys. Rev. B **55**, 1040 (1997).

[125] Y. E. Lozovik and V. I. Yudson, Sov. Phys. JETP **44**, 389 (1976).

[126] K. Moon, H. Mori, K. Yang, S. M. Girvin, A. H. MacDonald, L. Zheng, D. Yoshioka, and S. C. Zhang, Phys. Rev. B **51**, 5138 (1995).

[127] P. P. Ruden and Z. Wu, Appl. Phys. Lett. **59**, 2165 (1991).

[128] L. Świerkowski, D. Neilson, and J. Szymański, Phys. Rev. Lett. **67**, 240 (1991).

[129] D. Neilson, L. Świerkowski, J. Szymański, and L. Liu, Phys. Rev. Lett. **71**, 4035 (1993).

[130] M. Alatalo, P. Pietiläinen, T. Chakraborty, and M. A. Salmi, Phys. Rev. B **49**, 8277 (1994).

[131] S. Das Sarma and P. I. Tamborenea, Phys. Rev. Lett. **73**, 1971 (1994).

[132] F. Rapisarda, *Phase Diagram of coupled electron layers*, Ph.D. thesis, University of Trieste, 1996.

[133] A. H. MacDonald, Phys. Rev. B **37**, 4792 (1988).

[134] S. Shapira, U. Sivan, P. Solomon, E. Buchstab, M. Tischler, and G. B. Yoseph, Phys. Rev. Lett. **77**, 3181 (1996).

[135] Y. Katayama, D. C. Tsui, H. C. Manoharan, and M. Shayegan, Surf. Sci. **305**, 405 (1994).

[136] X. Ying, S. R. Parihar, H. C. Manoharan, and M. Shayegan, Phys. Rev. B **52**, R11611 (1995).

[137] Y. Katayama, D. C. Tsui, H. C. Manoharan, S. Parihar, and M. Shayegan, Phys. Rev. B **52**, 14817 (1995).

[138] S. Conti and G. Senatore, Europhys. Lett. **36**, 695 (1996).

[139] Y. Katayama and D. C. Tsui, Appl. Phys. Lett. **62**, 2563 (1993).

[140] L. Zheng and A. MacDonald, Phys. Rev B **49**, 552 (1994).

[141] J. P. Eisenstein, L. N. Pfeiffer, and K. W. West, Phys. Rev. B **50**, 1760 (1994).

[142] R. E. Prange and S. M. Girvin, *The Quantum Hall Effect, Graduate texts in contemporary physics*, 2nd ed. (Springer-Verlag, Berlin, 1990).

[143] T. Chakraborty and P. Pietiläinen, *The Quantum Hall Effects*, 2nd ed. (Springer, New York, 1996).

[144] S. D. Sarma and A. Pinczuk, *Perspectives in Quantum Hall Effects* (Wiley, New York, 1996).

[145] R. B. Laughlin, Phys. Rev. B **23**, 5632 (1981).

[146] B. I. Halperin, Phys. Rev. B 2185 (1982).

[147] A. H. MacDonald and P. Streda, Phys. Rev. B **29**, 1616 (1984).

[148] C. W. Beenaker and H. van Houten, in *Solid State Physics*, edited by H. Ehrenreich and D. Turnbull (Academic, New York, 1991), Vol. 44, p. 1.

[149] X. G. Wen, Phys. Rev. B **41**, 12838 (1990).

[150] X. G. Wen, Phys. Rev. B **44**, 5708 (1991).

[151] X. G. Wen, Int. Journ. Mod. Phys. B **6**, 1711 (1992).

[152] J. J. Palacios and A. H. MacDonald, Phys. Rev. Lett. **76**, 118 (1996).

[153] F. P. Milliken, C. P. Umbach, and R. A. Webb, Solid State Comm. **97**, 309 (1996).

[154] A. M. Chang, L. N. Pfeiffer, and K. W. West, Phys. Rev. Lett. **77**, 2538 (1996).

[155] A. V. Shytov, L. S. Levitov, and B. I. Halperin, Phys. Rev. Lett. **80**, 141 (1998).

[156] M. Grayson, D. C. Tsui, L. N. Pfeiffer, K. W. West, and A. M. Chang, Phys. Rev. Lett. **80**, 1062 (1998).

[157] U. Zülicke and A. H. MacDonald, Phys. Rev. B **54**, 16813 (1996).

[158] C. de Chamon and X. G. Wen, Phys. Rev. B **49**, 8227 (1994).

[159] C. W. L. Beenakker, Phys. Rev. Lett. **64**, 216 (1990).

[160] A. M. Chang, Solid State Comm. **74**, 871 (1990).

[161] D. B. Chklovskii, B. I. Shklovskii, and L. I. Glazman, Phys. Rev. B **46**, 4026 (1992).

[162] M. Ferconi, M. R. Geller, and G. Vignale, Phys. Rev. B **52**, 16357 (1995).

[163] O. Heinonen, M. I. Lubin, and M. D. Johnson, Phys. Rev. Lett. **75**, 4110 (1995).

[164] L. Brey, Phys. Rev. B **50**, 11861 (1994).

[165] D. B. Chklovskii, Phys. Rev. B **51**, 9895 (1995).

Elenco delle Tesi di perfezionamento della Classe di Scienze
pubblicate dall'Anno Accademico 1992/93

"CompoMat" Loc. Braccone, 02040 Configni (RI), Italy
Finito di stampare per conto della "CompoMat" dalla Nuova Grafica 86 nel gennaio 2000